ALSO BY PHILLIP HOOSE

*Unbeatable: How Crispus Attucks Basketball Broke
Racial Barriers and Jolted the World*

*The Boys Who Challenged Hitler:
Knud Pedersen and the Churchill Club*

The Race to Save the Lord God Bird

Moonbird: A Year on the Wind with the Great Survivor B95

Claudette Colvin: Twice Toward Justice

Perfect, Once Removed: When Baseball Was All the World to Me

We Were There, Too!: Young People in U.S. History

Hey, Little Ant (with Hannah Hoose)

*It's Our World, Too!: Young People Who Are Making a
Difference—How They Do It—How You Can, Too!*

Necessities: Racial Barriers in American Sports

Hoosiers: The Fabulous Basketball Life of Indiana

*Building an Ark: Tools for the Preservation of
Natural Diversity Through Land Protection*

DUET

*Our Journey in Song with the
Northern Mockingbird*

DUET

*Our Journey in Song with the
Northern Mockingbird*

PHILLIP HOOSE

FARRAR STRAUS GIROUX
NEW YORK

Farrar Straus Giroux Books for Young Readers
An imprint of Macmillan Publishing Group, LLC
120 Broadway, New York, NY 10271 • fiercereads.com

Our books may be purchased in bulk for promotional, educational, or business
use. Please contact your local bookseller or the Macmillan Corporate and
Premium Sales Department at (800) 221-7945 ext. 5442 or by email at
MacmillanSpecialMarkets@macmillan.com.

Library of Congress Cataloging-in-Publication Data
Names: Hoose, Phillip M., 1947– author.
Title: Duet : our journey in song with the northern mockingbird / Phillip Hoose.
Description: First edition. | New York: Farrar, Straus and Giroux Books for Young
Readers, 2022. | Includes bibliographical references and index. | Audience: Ages 12–18
| Audience: Grades 7–9 | Summary: "The story of the impactful partnership between
humans and mockingbirds—a case study in courage, resiliency, and the power of a
song—written by nonfiction powerhouse Phil Hoose" —Provided by publisher.
Identifiers: LCCN 2021046012 | ISBN 9780374388775 (hardcover)
Subjects: LCSH: Mockingbirds—Juvenile literature. | Birdsongs—History—Juvenile
literature. | Human-animal relationships—Juvenile literature.
Classification: LCC QL696.P25 H66 2021 | DDC 598.8/44—dc23
LC record available at https://lccn.loc.gov/2021046012

First edition, 2022
Book design by Trisha Previte
Printed in China by Toppan Leefung Printing Ltd.,
Dongguan City, Guangdong Province

ISBN 978-0-374-38877-5 (hardcover)
1 3 5 7 9 10 8 6 4 2

To Charles Duncan, Ben Gregg,
William Housty, Larry Master, Margaret Ormes,
Keith Ouchley, and Carlos Peña

The most adventuresome of my birding companions, with thanks

TABLE OF CONTENTS

INTRODUCTION: What's That Sound?   3

CHAPTER ONE   11

Four Hundred Tongues: *Mockingbirds and Native Americans*

CHAPTER TWO   17

Settlers, Explorers, and Mock-Birds: *1492 to 1750*

CHAPTER THREE   25

"Dick Sings": *Thomas Jefferson and the First White House Pet, 1803*

CHAPTER FOUR   33

Mocker v. Nightingale—Full-Throated Rivalry: *Colonial Period*

CHAPTER FIVE   39

A Rattlesnake Climbs a Tree: *Audubon, Wilson, and* Plate 21, *1827*

CHAPTER SIX   47

Turning Point: *Charles Darwin and the Galápagos Mockingbirds, 1835*

CHAPTER SEVEN   57

Listening to the Mockingbird: *Septimus Winner and
Richard Milburn, 1855*

CHAPTER EIGHT   63

A Visit to High Street Market: *Philadelphia, Mid-19th Century*

CHAPTER NINE   69

Parrott Shells and Minié Balls: *The Mocker in the Civil War,
1861 to 1865*

CHAPTER TEN 75
Women to the Rescue: *1865 to 1918*

CHAPTER ELEVEN 83
Hush, Little Baby: *1918 to 1937*

CHAPTER TWELVE 89
Amelia Laskey, Citizen Scientist: *1920s to 1970s*

CHAPTER THIRTEEN 99
To Kill a Mockingbird: *1960*

CHAPTER FOURTEEN 103
Improvising: *Mockers on the Move, 20th and 21st Centuries*

CHAPTER FIFTEEN 111
"They Know Me!": *The Genius of Mockingbirds, 2007*

EPILOGUE 117

**WHAT YOU CAN DO TO HELP MOCKINGBIRDS
AND OTHER SONGBIRDS** 123

ACKNOWLEDGMENTS 126

SOURCE NOTES 128

INDEX 137

DUET

*Our Journey in Song with the
Northern Mockingbird*

(Alamy)

What's That Sound?

*Some humans can go beyond simple speech and train their voices
to sing opera or other types of music. . . . Mockingbirds might be
doing something similar.*

—William Young, *The Fascination of Birds*

Y ou're walking down a city sidewalk, lost in thought, when a song
erupts from high above you. It's a loud, bright, liquid stream of
musical notes and phrases. You shield your eyes and try to see where
it's coming from, but you can't find it at first. Then, there it is . . . you
spot a single, slender gray bird about the size of a robin bellowing from
the top of a telephone pole. Its head is thrown back, its bill is wide
open, and its total body rattles with the force of this performance. Its
long, sleek tail is cocked up sharply, giving the bird the silhouette of a
checkmark. As you listen more carefully, you realize it isn't just singing one song; it's sampling song fragments one after another, like a DJ
pulling material together. The bird is improvising.

You look around to see if anyone else is noticing. No one is even
looking up, although our soloist seems to be demanding attention.

(Will Elder / NPS)

Sure enough, attention comes. A red-tailed hawk, a winged monster, sweeps into view, talons extended, picking up speed, bearing down on our singer. With perfect timing, the gray bird waits, and waits, and then rises into the air just as the hawk slides beneath him. Now the battle is on. The singer seems to skid in the air, reverses course, and, scolding, takes off after the hawk, a bird several times its size. The gray bird is agile and fierce, strafing the hawk again and again. At one point, the gray bird jumps aboard the monster's back, where it pecks and tears at feathers until it lets go and the wounded giant flaps away. The gray bird returns to its telephone pole, smooths its feathers, and resumes its performance, as if nothing has happened.

Our pugnacious, hawk-attacking bird is the northern mockingbird. The word *mockingbird* applies to seventeen species of songbirds from the Mimidae family of mimic thrushes. The northern mockingbird's scientific name is *Mimus polyglottos*, a Latin term that means "many-tongued mimic" in English—in honor of its ability to imitate the songs of other birds.

"Mockers," as they are often called, are widely known for their spectacular song. They can imitate just about anything that makes a sound—police sirens, telephones, cats, alarm clocks, dogs, and human songs. But the mocker's primary set list consists mostly of the songs of other birds, learned mainly from the birds themselves. Mockers can learn two hundred or more songs in their lifetimes.

Some people rank the northern mockingbird as the greatest singer in the animal kingdom. Northern mockers can sing for hours without stopping, day or night. In springtime, male mockers sing to impress females and defend nesting territory. Females sing, too, at other times of the year, but in spring they are silent judges, listening carefully to auditioning males. A long and varied set list, delivered with skill and confidence, and with high volume, wins the day.

What is the advantage of knowing and performing hundreds of songs? Ask any rock star. Research shows that as a male mockingbird expands his repertoire, he becomes increasingly attractive to females. The experienced male proclaims, through his brilliant song, "I've been around. I'm a survivor. I've already found us a nesting territory with plenty of food. I've driven away the competition. Why settle for silver when you can have gold?"

The secret to the mockingbird's song is an organ, sunk deeply in the bird's chest, called a syrinx. It's the songbird's voice box, found in more than half of all the world's eight thousand bird species. Though not much bigger than a raindrop, the syrinx efficiently uses nearly all the air that passes through it. By contrast, a human uses only 2 percent of the air exhaled through our larynx to make sounds. Mockers have seven pairs of syringeal muscles, allowing them to deliver complex songs

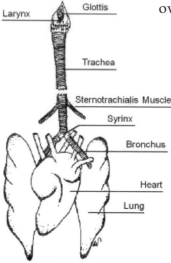

Larynx
Glottis
Trachea
Sternotrachialis Muscle
Syrinx
Bronchus
Heart
Lung

Diagram showing the location of a songbird's syrinx (Wikimedia)

over and over with seeming ease. "There is nothing that looks like a syrinx in any related animal group in vertebrates," observed Denis Duboule, a geneticist at the University of Geneva.

A recent survey led by Cornell University has found that there are almost three billion fewer birds in North America today than there were a mere fifty years ago. We have lost almost one third of our birds since 1970, much of it due to loss of habitat—the places equipped with the food, shelter, and nesting areas a particular bird species needs—as a result of climate change. Collisions with window glass, hunting by cats, and the use of pesticides were named by the study as other major culprits.

And yet, while the overall bird population has gone down, the northern mockingbird has actually *expanded* its range. At a time when

Breeding
Nonbreeding (scarce)
Year-round

A northern mockingbird's range map (from eBird.org, published by the Cornell Lab of Ornithology)

environmental stresses are causing massive bird die-offs throughout the world, the northern mockingbird continues north, spreading throughout most of the United States and on into Canada. It can live within a wide range of habitats, from parched deserts to suburban shrubs to city lawns.

Scientists using DNA techniques conclude that mockingbirds may have evolved from a common starling-like ancestor more than twenty million years ago. By the time the first humans found their way to what is now known as the United States, perhaps fifteen thousand years ago, mockingbirds were well established from coast to coast.

Climate scientists have developed a saying: Adapt or die. It means that those species that can't quickly fit in to altered habitats will struggle to survive. But mockingbirds, with their ability to adjust to a wide variety of habitats, have, in fact, made it through the millennia.

Over thousands of years of their survival, mockers have learned to live with us—*Homo sapiens.* They adapt easily to new foods and have learned to survive in harsh weather. Some mockingbirds migrate south in the winter, but usually only until food is readily available. When the ground freezes and insects burrow beneath the ice, most mockers fluff up their feathers, find shelter, switch to a diet of berries, and tough it out. How tough? There is a record of one mockingbird surviving a Canadian winter in the province of Alberta when temperatures plummeted to −24°C (−11°F). But mockers are also on your front lawn, singing from telephone poles, hedges, wires, and rooftops in towns, suburbs, backyards, and parks, all year round. Mockingbirds are holding their own as the earth warms because they have developed a special awareness of the living creatures around them—including us. It has allowed them to adapt . . . and survive. For millennia they have built their nests near our dwellings and followed our movements, most recently abandoning farm and field to become denizens of suburbs and cities.

Our road together has not been bump free. Mockers are often noisy singers, keeping us up at night, as well as fierce protectors of their young, sometimes at human expense. And if mockers could speak, they would lodge complaints of their own: We humans have hunted them, sold them, and at one point, nearly driven them to extinction. At the heart of our connection is a bargain: Human life provides sustenance for the mockers' survival, and mockers provide us with inspiration and entertainment. Mockers get to dine on the fruits and berries that people have planted, as well as the insects that abound in cleared fields and watered lawns. And we humans get to hear and make music with none other than the King of Song himself, as John James Audubon referred to the mocker. Humans and mockingbirds are a duet—two very musical species united in song.

In 1784, Ben Franklin wrote a letter to his daughter expressing disappointment that our new country, the United States of America, had adopted the bald eagle as its national symbol. Franklin complained that the bald eagle was a "rank coward" who merely stole food from other birds. Franklin preferred the wild turkey, "a bird of courage." Bird expert David Sibley confirmed the turkey's high reputation more than two hundred years later in his book *What It's Like to Be a Bird*, saying, "No other North American bird has such a complex history with humans."

In these pages, we nominate a third candidate. We suggest that no bird has meant as much to American life as the northern mockingbird. We contend that the mocker is the bird species that has been most tightly connected to human myth, song, literature, entertainment, art, business, battle, and science. And sleep. Mockers have been our

A mockingbird keeps a close eye on a bald eagle. (© Brad Dinerman)

inseparable companions for thousands of years, and their torrential song still seems as amazing today as it must have a thousand years ago.

The great Chilean poet Pablo Neruda expressed his wonderment of songbirds as a question in his poem "Ode to Bird-Watching":

How
Out of its throat
Smaller than a finger
Can there fall the waters
Of its song?

(Theodore Emery)

Four Hundred Tongues

Mockingbirds and Native Americans

[We] send greetings and thanks to all the Animal life in the world.
They have many things to teach us as people . . . We are glad they
are still here and we hope that it will always be so.

—Haudenosaunee Thanksgiving Address

When Spanish explorers arrived in North America and claimed what they called the "New World," millions of Indigenous people were already living there. They were scattered in nations and bands and tribes throughout the chain of continents and islands that stretches from Alaska to Chile. Mockingbirds could be heard throughout much of that territory.

The mockingbird's song captured the attention and respect of Indigenous Americans, and plays a central role in many Native American myths and legends that explain how life began or, especially, how languages developed. A few examples:

Cherokees called the mocker *Cencontlatolly*, which translates to "four hundred tongues." They fed their children mockingbird heads to increase their intelligence.

In Shasta Indian mythology, the mockingbird watches over the dead. To the Maricopa, one who dreams of a mockingbird is endowed with special powers. The mockingbird is a wise mediator who settles disputes in O'odham folklore.

In the creation stories of the Hopi and other Pueblo tribes, it was Mockingbird—the most gifted of all linguists—who taught the people to speak. Mockingbirds could even pass messages between people and gods.

In the Pueblo story, animals and Pueblo people initially lived together underground. Both groups sought to rise to live aboveground, but they didn't know the way. Badger and Shrike—a songbird—were advance scouts who finally found the way out. They received a welcoming promise of land for the Pueblo people. Five days later, the first Pueblos emerged from the lower world through the mouth of a cave. As they did, Mockingbird, sitting just outside the cave entrance, gave them each a language.

A story from Mayan culture explains how the mockingbird became the best singer. It is the story of X-chol-col-chek, a young female mockingbird who agrees to use her great singing talent to help a friend. The friend, a young female cardinal, has been ordered by her father to sing to the entire community. But the cardinal is terrified because she knows she is a poor singer. She talks her mockingbird friend into hiding inside a hollow tree and singing in her place.

But just before the concert the cardinal father finds out about the switch. Instead of punishing his daughter, he invites X-chol-col-chek to come out from her hiding place and sing for the crowd. The small, gray bird trembles with fright at first, but then gains confidence as the audience responds to her brilliant performance. Wings flap in wild applause. She takes many bows. From that time on, all her mocking-bird descendants inherit her lovely voice. That is how the mocking-bird became the best singer. But, goes the story, cardinals never have learned to sing very well.

Another Hopi story offers a very different explanation for how Mockingbird gave their people many languages. This happened at a time when heavy rain and punishing winds sent water rising over riverbanks and drove the people from their homes. The downpour showed no sign of stopping. Stranded people began to carp at one another. The chief worried that some people were secretly forming plans to return to their flooded homeland despite the danger.

He called a council and asked his wisest leaders what could be done to preserve the peace. One counselor suggested there were too many people who spoke the same language. It was too easy for them to plot and scheme. Why not split the tongues?

"How could we do this?" the chief asked.

"Well," he replied, "don't we have Mockingbird, who knows many songs? Why shouldn't he give us many different languages?"

The great vocalist sang from inside a tree to help a friend. (Shutterstock)

The chief was intrigued. It seemed worth a try. They prepared an offering for Mockingbird and set it out. They started singing Mockingbird's calling songs and, after four songs, a slim, gray, long-legged bird with its tail held high fluttered down before them. The chief stated his request, and Mockingbird said yes, he could scramble the language. It would be done by the following morning. The bird asked the chief if he would like to speak a different language, but the chief said he would rather keep his own.

When the people awoke the next morning, they couldn't understand one another. Baffled and upset, they went to the chief, but they couldn't understand him either. The mockingbird, adept at languages, had to interpret. And still does.

As time has passed, new mockingbird tales have arisen. A story told by Mexican-American folklorist Jovita González explains why the mockingbird has a white stripe on each wing: At one time, all animals spoke the same language. Mockingbird became so convinced that his sound was the best of all that he became conceited. He boasted to his wife that the next day he would perform a concert to the flowers and cause them to dance. "Con el favor de Dios," his wife said: "If God wills it."

The next day, as Mockingbird cleared his throat to sing from the top of a sweet acacia tree, he was snatched away by a hawk. As he was carried up high in the air, he realized his vanity. He repented. "O God, it is you who make the flowers bloom and the birds sing, not I." The hawk released him and he fell to the ground, landing heavily. He was comforted by a white dove who had a nest nearby. "My wings!" said the mockingbird. "How tattered and torn they look!" The white dove, taking pity on him, plucked three feathers from her own white wings to repair the mockingbird's torn wings. And to this day, the

The mockingbird flashes the white feathered stripes on its wings. (Ruth Hoyt)

mockingbird bears the dove's bright stripe on each wing as a reminder not to be so proud.

For many years, these and countless other stories and songs have become polished like smooth stones in the telling and retelling. And Mockingbird, that brilliant communicator who speaks in four hundred tongues and shuttles messages to and from the gods, has often been a leading character.

(*Lang Elliott*)

CHAPTER TWO

Settlers, Explorers, and Mock-Birds

1492 to 1750

He imitateth . . . at all hours in the night.

—Thomas Glover, *An Account of Virginia*

Christopher Columbus captained three wooden ships across the Atlantic Ocean, from Spain to an island in the Caribbean Sea, arriving in October of 1492. The Spaniards thought they had discovered a back route to India, but it didn't look or sound as they had expected. For the next five months, they explored the land and waters of the warm Caribbean, particularly the islands of Juana (now known as Cuba) and Hispaniola (the Dominican Republic and Haiti). Ever on guard for an ambush, the Spanish sailors clanked through the blazing heat in heavy armor.

Columbus kept a journal intended to impress the Spanish royals, Queen Isabella and King Ferdinand, who had paid for most of the voyage. He would present the journal to them when he returned to Spain. He lavishly described the newly discovered "India" as a "cornucopia," or as Belgian artist Ana Torfs interpreted his writings, a "paradise of wondrous flowers, with a thousand variety of trees and remarkable fruits, not to mention astonishing fish and birds of the most dazzling colors."

Again and again, Columbus reported the presence of the nightingale, the small bird whose complex and beautiful song made it the poetic soul of Europe, and one of the queen's favorite songbirds. Early in the visit, Columbus wrote that he "went a short distance into that country . . . and heard sing the nightingale and other songbirds like those of Castile [Spain]." Months later he repeated, "The nightingale and other small birds of a thousand kinds were singing in the month of November when I was there."

But the birds the Spanish explorers heard were not nightingales. There were no nightingales in "the New World." The nearest nightingales were back in Europe, an ocean behind them. Columbus's crew members were probably hearing northern mockingbirds. As Cuban biologist Carlos Peña says, "I suspect [Columbus] was referring to the northern mockingbird, because it is abundant in different environments in Cuba and because it sings loudly from exposed perches." Maybe Columbus assumed that only a nightingale could sing such a spectacular song, or maybe he was just inventing the sightings to please Her Majesty. The two birds do not look much alike—the mockingbird is slender and gray, and the nightingale compact and brown.

Ana Torfs has studied Columbus's journal in detail. She found that *nightingale* was one of the words he used the most, along with *sign, gold, tree, cross, believe, trade, parrot, danger, wonder,* and *weapon.* Her conclusion: "When we bump up against the limits of our imagination and knowledge, reflections of what we already know become the blueprint for new descriptions and definitions." In other words, the nightingale *should* be there, it *must* be there, so Columbus *puts* it there.

Torfs saw Columbus as one who looked but didn't see, or saw only what he wanted to.

In the century that followed, more and more Spanish adventurers explored Central and South America, Mexico, and the Caribbean, establishing settlements and missions.

England defeated the Spanish navy in 1588, clearing the way for British exploration and settlement in the New World. Settlers and mockers became a duet—a musical partnership—adding to the songs and chants created long before by Native Americans. In 1705, historian Robert Beverley sized up the musical bond between mockingbirds and colonists: *"Mock-birds . . . love Society so well, that whenever they see Mankind, they will perch upon a Twigg very near them, and sing the sweetest wild Airs in the World: But what is most remarkable in*

(*Alamy*)

these *Melodious Animals, they will frequently fly at small distances before a Traveller, warbling out their Notes several Miles an end, and by their Musick make a Man forget the Fatigues of his Journey.*"

Not everyone was charmed. Early explorers complained in their journals of the mocker's annoying ability to sing through the night and keep them awake. In 1676, Thomas Glover, perhaps rubbing his eyes as he wrote, complained about the mockingbird in his *Account of Virginia*: "He imitateth all the birds in the woods . . . he singeth not only in the day, but also at all hours in the night."

Mark Catesby's "Mock-Bird"

Mark Catesby gave many birds their first English-language names. He tended to use multiword descriptive names, such as "blew grosbeak," "yellow-breasted chat," and "painted finch." The "mock-bird," as Catesby called the mocker, was later named by Carl Linnaeus in his 1758 tenth edition of *Systema Naturae*—a framework for classifying and naming all plants and animals in the world—*Turdus polyglottos.*

Two of America's earliest and most important explorers were surveyor John Lawson and naturalist and artist Mark Catesby. Setting out in 1700, Lawson explored by canoe the barrier islands off the South Carolina coast and then hiked to the mountaintops of North Carolina. He described the plants, animals, and Indigenous people he encountered in a book titled *A New Voyage to Carolina*. Lawson wrote that mockers "are held to be Choristers of America, as indeed they are. They sing with the greatest Diversity of Notes, that is possible for a Bird to change to."

In 1712, Mark Catesby journeyed from England to America to paint and study the life forms present in the New World. He hired Native Americans to guide him through lands that had not been explored by Europeans. Catesby produced the first book describing the plants and animals of what is now the southeastern United States. The book included paintings of 220 species, including the songster he called the "mock-bird."

Catesby's portrait of the "mock-bird" (Alamy)

As the land began to fill with settlers and enslaved people kidnapped from Africa to perform backbreaking labor, the mockingbird didn't retreat into the wilderness. Instead, if anything, the bond with humans became tighter. Mockers nested in the shadows of crude cabins and great plantation mansions alike. They sang from rail fences and farm hedges and chimney tops. They built their nests in shrubs and flitted among the bearded limbs of moss-covered oak trees. In the Southwest, they nested in prickly cacti and dined on pale green sage plants.

They would devour just about anything, including quantities of worms, beetles, grubs, moths, caterpillars, and grasshoppers, feeding both on the ground and in trees. When fruits ripened, they took their full share, eating figs, grapes, strawberries, blackberries, dewberries, prickly pears, mulberries, and a far larger range of soft fruits and berries. Poison ivy berries—rich in fat and protein—were a special mockingbird treat. They were even observed drinking the sap from cuts on recently pruned or injured trees.

Mockingbirds were opportunists, birds that took chances, birds that weren't afraid of anyone or anything. Frontier children gathered to watch mockers turn somersaults in the air on moonlit nights.

At the heart of the mocker-human partnership, the duet, was an unspoken tradeoff: food for entertainment. Settlers planted fruit trees and berry bushes. The mockers gobbled down the food on the spot or carried it off to feed gape-mouthed nestlings.

In the years before radios and recorded songs, and before telephone wires crisscrossed the sky, America was a singing country. Many people could play instruments. Native Americans accompanied songs and chants with flutes and drums. Around the fire, pioneer families played fiddles, banjos, and harmonicas.

And the mocker took it all in, threw back its head, and claimed its place in the choir.

Working as a Family

When the female mocker chooses a mate, the couple works together to build a nest, usually in a hedge or a bush a few feet off the ground. The male does most of the construction work while the female keeps watch for predators. The female lays an average of four brown-splotched blue eggs, which hatch in about two weeks. Nestlings are born blind, helpless, and fuzz-coated. The nestlings must be fed by their parents for their first two weeks. Mockingbirds are famously attentive parents, attacking cats, dogs, and even humans that stray too close to their nests.

[Top and bottom] (Alamy)

Jefferson's White House study. Note the mockingbird flying free.
(Peter Waddell for the White House Historical Association)

CHAPTER THREE

"Dick Sings"

Thomas Jefferson and the First White House Pet, 1803

A superior being in the form of a bird.

—Thomas Jefferson

As of this writing, a total of nineteen US presidents have kept birds as pets. But the first pet ever to live in the White House was a sweet-singing and very smart northern mockingbird named Dick, who joined President Thomas Jefferson in the White House in 1803. The mocker was the president's pride and joy.

Jefferson had long admired mockingbirds. He bought his first mocker for five shillings in 1772 from an enslaved person owned by Jefferson's father-in-law, John Wayles. In those days, mockingbirds were captured and sold widely as pets. They sang from their cages and could be trained to imitate all sorts of sounds, including songs, musical instruments, and the squawks, whinnies, brays, quacks, and growls of farm animals. Some purchasers would only buy "singing" mockingbirds—birds caught and taught to sing a few songs before they were offered for sale. Other owners preferred to let the birds imitate the sounds they heard around their cages.

Jefferson bought another mocker the following year. As much as he enjoyed his new pets, Jefferson worried that mockers were becoming scarce. So many nestlings had been trapped along the tidal rivers of eastern Virginia and sold in city markets that Jefferson wondered if any free mockingbirds would ever reach Monticello, his hilltop home in central Virginia. He knew that mockingbirds followed humans, and human settlement had not yet reached Monticello. But Jefferson wondered if there would be wild mockers left by the time the land around Monticello was cleared for farms and fields.

But in 1793, Jefferson's son-in-law Thomas Randolph heard a mocker's unmistakable song spilling from the branches of a cedar tree at Monticello. Randolph immediately wrote to Jefferson, who was then living in Philadelphia. Jefferson, widely regarded as one of Virginia's leading bird experts, replied at once to his daughter Martha Jefferson Randolph. He made no effort to conceal his excitement: "I sincerely congratulate you on the arrival of the mocking-bird," he wrote. He also issued a warning: "Learn all the children to venerate it as a superior being in the form of a bird, or as a being which will haunt them if any harm is done to itself or its eggs."

Thomas Jefferson became the third US president on March 4, 1801. He reluctantly moved into a White House still under construction. At the time it was the largest residential house in America. Jefferson found it a dusty and depressing collection of empty rooms. He suffered from back pains and headaches. Financial woes and a sense of

Presidential Birds

Before the White House was built, President George Washington's wife, Martha, kept a caged parrot at their house in Mount Vernon.

President Andrew Jackson brought a parrot and taught it to swear—which it did at Jackson's funeral.

President Abraham Lincoln received a wild turkey as a gift in 1863. When he announced to his family that he planned to slaughter it for a Christmas feast, his son Tad burst into tears. Lincoln was surprised to learn that Tad had developed a powerful bond with the bird. He had named the turkey Jack, and it followed Tad everywhere he went. "He's a good turkey, and I don't want him killed," the boy pleaded. The president gave in, writing out a formal pardon for the bird.

In addition to Jefferson, Rutherford B. Hayes and Grover Cleveland kept mockingbirds as caged White House pets. But President Cleveland never warmed to life with a mockingbird. One night when Cleveland was up late working, his mocker's song repeatedly disrupted his concentration. He had the cage moved to a distant part of the White House, but now the silence worried him even more than the singing had. He found himself wondering if the mocker would catch cold in his new location. Mr. Pendel, his aide, spent that long night moving the cage from place to place, trying to find a spot out of the president's earshot where the bird could stay warm.

President Calvin Coolidge's wife owned a mockingbird as well, but when she learned that it was illegal in Washington, DC, to confine a mockingbird in a cage—punished by a five-dollar fine or a month in prison—she gave up her mocker. She explained, "I was reluctant to part with my chorister, but I was even more averse to embarrassing my country by the imprisonment of its First Lady."

personal loss deepened Jefferson's gloom. His wife, Martha, had died years before, and now he was separated from his beloved grandchildren. He was living in the White House with his aide Meriwether Lewis.

Jefferson spent much time in his study, an almost-empty vault with an eighteen-foot-high ceiling, a floor polished to a mirrorlike sheen, and a few pieces of furniture. Into the study he moved objects of personal interest—tables that were soon piled high with books and papers, his telescope, his maps, his volumes, his globe, his houseplants, and his violin.

Still the rooms felt empty and bare. He ordered wallpaper, wine, and furniture from France, but nothing could fill up the silent spaces or restore the cheer that he felt with his family at Monticello. And then an idea struck him: He remembered the mockingbirds he had owned when he was a much younger man. He loved those birds. What creature could possibly be a sunnier companion than a mockingbird? If mockingbirds had worked to cheer him up before, they would work again.

Jefferson purchased two mockers in 1803 for ten and fifteen dollars. They were "singing" birds, who had already learned a vocabulary of songs. Jefferson bought two more mockers the following year and soon added another. He kept track of them in his "weather memorandum" book.

1806 Jan 22. "N.O. [New Orleans] mockg. bird begins to sing"

1806 Feb 19. "2nd Mockg. bird sings."

1806 Feb 25. "The old bird begins to sing."

1808 Jan 23. "Orleans bird sings"

1808 Jan 31. "The old mock. bird sings."

1808 March 2. "The middle aged bird sings."

Finally, on March 3, 1808, Jefferson wrote, "Dick Sings."

At last, a name. Dick became Jefferson's favorite mocker, the only bird he gave a name. According to Jefferson's friend and early American historian Margaret Bayard Smith, Dick's cage dangled among the roses and geraniums on a windowsill in the president's study. Jefferson kept the cage door open. The presidential mockingbird spent its days whizzing around Jefferson's study, perching on one object or another to "regale him with its sweetest notes."

Dick sat on Jefferson's shoulder as he worked, and sometimes took his food from Jefferson's lips. At nap time, Dick would hop behind Jefferson up the stairs, and then perch on a couch and sing the president to sleep. "How he loved this bird!" Smith observed. "He could not live without something to love . . . his bird and his flowers became objects of tender care."

Captivity

For a man who wrote so eloquently about the ideals and principles of freedom, Jefferson also owned dozens of people and depended on them for his livelihood and comfort. Jefferson opposed slavery unless the slaves in question happened to be his own. He owned more than six hundred enslaved people in his lifetime. They cooked his food, kept his home, and tended his fields. Upon his death, Jefferson sold one hundred thirty slaves to pay down his massive debts. He freed only two.

Along those lines, some find it telling that Thomas Jefferson could worry about the status of wild bird populations while finding personal enjoyment from caged birds.

When Jefferson left the White House in the spring of 1809 and moved back to his beloved Monticello, he sent a letter to a friend about a matter that brought him great relief. "My birds arrived here in safety," he wrote, "and are the delight of every hour."

*A photo artist's tribute to the acrobatic
mockingbird (Edward Rooks)*

CHAPTER FOUR

Mocker v. Nightingale—
Full-Throated Rivalry
Colonial Period

It may not be improper here to consider, whether the nightingale may not have a very formidable competitor in the American mockingbird. . . . During the space of a minute, he imitated the woodlark, chaffinch, blackbird, thrush, and sparrow. . . . His pipe comes nearest to our nightingale of any bird yet met with.

—British naturalist Daines Barrington, in the scientific journal *Philosophical Transactions*, 1773

During the years leading up to the American Revolution, when the area was still a collection of British colonies, the nightingale (*Luscinia megarhynchos*), a European songbird, was widely assumed to be the best singer in all of nature. The nightingale is a small, chunky brown bird that escapes attention until it opens its throat to sing. And the song is truly special—a tuneful jumble of whistles, trills, buzzes, and gurgles, often delivered at night.

For many centuries, the nightingale's song has inspired European fairy tales, operas, books, and a great deal of poetry, some of it time-less. Homer wrote about the nightingale in *The Odyssey*; Sophocles, Virgil, Geoffrey Chaucer, Shakespeare, and T. S. Eliot all also wrote about nightingales. Back in 1492, Christopher Columbus had falsely assured Queen Isabella that nightingales were singing from practically every branch in the New World. But early naturalists like Catesby never found them.

There was, however, another song that immediately caught the attention of those naturalists. It came from a bird that made almost no attempt to conceal itself from human company. It seemed to crave attention. It was a slim gray songbird entirely absent from the Old World. But the *song*. It could sing loudly for hours without resting. The notes that poured forth were bright and true, some of them seem-ing to mimic the songs of other birds. Colonists found themselves comparing the Old World nightingale and this New World mock-bird, as Catesby had named it. Some found themselves thinking an almost seditious thought—that the mocker's song was just as special as the nightingale's tune. Maybe, some whispered—perhaps glancing around first—maybe even *more* impressive.

Soon a full-throated transoceanic rivalry blossomed between advo-cates of the nightingale and those of the mockingbird. Who had the better song? Thomas Jefferson wrote a goading letter to his friend and staunch nightingale booster, Abigail Adams: "In America [the night-ingale] would be deemed a bird of the third rank only, our mocking bird, and fox-colored thrush [the brown thrasher, another New World bird with a great song] being unquestionably superior to it." Years later, American bird painter John James Audubon agreed: "Some . . . persons have described the notes of the Nightingale as occasionally fully equal to those of our bird. . . . But to compare her essays to the finished talent of the Mocking Bird, is, in my opinion, quite absurd."

Attempts to establish a breeding population of nightingales in the British colonies failed. So did attempts by poets to capture the spirit of a mockingbird by using the same poetic techniques they had used to describe the nightingale.

Nightingale: a big sound from a little bird (Alamy)

The nightingale in a British poem was often a mournful figure who sang melancholy songs from deep, forested haunts, often at night. By contrast, the extroverted mockingbird woke up with a bang, sizzling with energy and looking for an audience through a flashing yellow eye.

Many years later, the Eastern Tennessee University ornithologist Dr. Fred Alsop offered an admiring description of the mockingbird. "It's an outgoing bird," he observed. "It puts a lot of effort into its displays, so it will often not just sit on the perch and sing, but it will fly up into the air, somersault back down, and be vocalizing the whole time. . . . I can see someone looking at this bird and getting excited."

The debate was at full throttle in 1892, when American poet Maurice Thompson compared the two birds in his poem "To an English Nightingale." Of the nightingale he wrote:

Dream on, O nightingale!
Old things shall fade and fail,
And the glory of the past shall not avail

And of the mockingbird:

Mine is the voice of Spring
My home is the land of the new,
And every note I sing
Is fresh as the morning dew;
For I am Freedom's bird

The mockingbird's song was the sound of a new nation.

Audubon's Plate 21 (National Gallery)

CHAPTER FIVE

A Rattlesnake Climbs a Tree
Audubon, Wilson, and Plate 21, *1827*

The extent of its compass, the great brilliancy of execution, are unrivaled. There is probably no bird in the world that possesses all the musical qualifications of this king of song.

—John James Audubon, *The Birds of America*

B
ird artists plunged into the wilderness in the early years of the new United States of America. Most were white men new to the lands of North America, and they were determined to portray the birds of this nation as they began their own exploration of it. In the years before lightweight cameras, zoom binoculars, or collapsible spotting scopes, the first bird painters shot the birds and painted them as specimens.

Gunpowder was as much a tool of their trade as paint. Alexander Wilson, renowned as the best of all bird painters at the beginning of the nineteenth century, typically entered the wilderness on horseback. He had a loaded pistol in each pocket, a rifle strapped across his shoulder, and a pound of gunpowder in a flask around his waist.

He stored five additional pounds of gunpowder in his saddlebags. When a bird aroused his attention, he leveled his shotgun, aimed, and squeezed off a hailstorm of round steel pellets known as birdshot, intended to stun or kill the feathered target without tearing up its plumage. Then he dismounted and collected the carcass and skinned it, removing the internal organs so they wouldn't rot, then packed it in salt and stuffed it full of cotton to preserve it until he would have time to paint it.

Wilson and the other bird artists were out to introduce settlers and pioneers to the birds of their vast new home. His plan was to combine the portraits into book form—folios, the books were called—and sell them. He funded his work by selling subscriptions—accepting money up front and sending illustrations in installments. Wilson was proud to say that President Jefferson had subscribed, as had each member of Jefferson's cabinet. Wilson basked in the warm glow of his unchallenged reputation as "the Father of Ornithology."

All that came crashing down on a Friday morning some years later. On March 9, 1810, Wilson showed up unannounced and unexpected at the general store in Louisville, Kentucky, owned by John James Audubon and his wife, Lucy. Wilson's intention was to quickly unroll a few paintings, dazzle Audubon with his work, sign him up for a subscription, and move on.

The Audubon that Alexander met that day was twenty-five years old and profoundly bored. He felt trapped. He hated every minute of running a store. Nothing he could do behind a counter matched the thrill of exploring the wilderness dressed in buckskin and with his hair tumbling down over his shoulders. His passion was shooting birds and painting the specimens. Deeply frustrated, Audubon couldn't seem to figure out a way to earn a living that would also make him happy.

That is, until the day Alexander Wilson came to call. What Wilson didn't know was that Audubon was also a bird artist, having completed nearly two hundred pieces. In Audubon's counting room, the two painters spread out their work and bent over the images. Probably through forced smiles, both men recognized that Audubon's were better.

Audubon had developed a completely original style. He arranged a specimen's limbs and feathers into poses on a large wire grid in order to tell dramatic stories. Audubon could adjust the poses almost like a puppeteer and hold the poses in place with pins. He worked up exciting scenes that showed the bird species in action: building nests, defending their young, or snatching up prey. Wilson's traditional poses looked stiff by comparison.

That afternoon, Audubon turned down Wilson's request for a subscription, which led to hard feelings that never healed. Instead, Audubon resolved to do the same thing Wilson was doing—create books filled with paintings of all the birds of America—only he would do it better. It took almost ten years, but in the fall of 1820, Audubon left Lucy behind to raise their two sons alone, and stepped aboard a flatboat in Louisville. For the next sixteen months, he traveled the Ohio and Mississippi Rivers, stepping ashore to shoot and paint bird specimens.

Mockingbirds were his constant companions. He was especially impressed by the mocker's courage. Audubon recognized, as would many bird observers for decades to come, that mockers sensed predators at the earliest moments and didn't hesitate to attack birds several times their size in defense of the nest. "Few Hawks attack the Mocking Birds," Audubon wrote, "as on their approach, however sudden it may be, they are always ready not only to defend themselves vigorously and with undaunted courage, but to meet the aggressor half way, and force him to abandon his intention."

Audubon was especially impressed by the mockingbird's courage.
(Nick Dunlop Photography)

The mockingbirds' spectacular songs made them too valuable to shoot, as he might have shot other specimens. Instead, collectors—many of them young boys—reached into mocker nests and plucked out the blind, tiny birds and handed them over to adults, who paid for them. The new "owners" in turn stuffed the birds in cages and fed them until they were old enough to carry to market and sell as songbirds.

This piracy didn't seem to bother Audubon at all; in fact, he offered practical advice to those who captured and sold newborn mocking-birds. "The Mocking Bird is easily reared by hand from the nest," he wrote, "from which it ought to be removed when eight or ten days old. It becomes so very familiar and affectionate, that it will often follow its owner about the house."

Audubon's talent blossomed, and his artistic reputation grew steadily. He adopted Wilson's pay-as-you-go subscription model. Every few months, subscribers would receive a package of five prints:

one large bird, one medium-sized bird, and three smaller birds. When he was finished, he had painted 435 plates depicting a total of 1,055 birds. The entire collection was titled *Birds of America*.

His work was admired, but controversial. Some of his harshest critics were Wilson's friends, who harbored hard feelings toward Audubon even after Wilson died in 1813.

Other bird artists—perhaps jealous—accused Audubon of inventing dramatic scenes that could never happen in nature. Why wasn't it enough, they complained, for Audubon to paint a red-shouldered hawk flying or perched on a limb? Audubon *had* to have the bird crashing into a family of horrified and panic-stricken quail.

How Many Species?

Before the twentieth century and into the early 1900s, many new bird species were discovered throughout the United States and were named by explorers and scientists. As collecting expeditions penetrated remote, species-rich areas of North America, up to several hundred new species per decade were named, painted, and described. These bird types were already well known by local Native Americans.

Early bird artists contributed greatly to our knowledge of new species. Wilson alone named twenty-six bird species, a few of which he named after himself. He wrote 314 species accounts and provided illustrations for all of them. Audubon is credited with naming twenty-five new bird species.

The painting that sent Audubon's critics howling in protest was the portrait of a mockingbird family. In 1821, Audubon collected several mockingbird specimens in Louisiana and attached their limbs to his wire grid, positioning the birds to support the dramatic story he wanted to tell. Audubon's *Plate Number 21*, as he called it, was not published until 1827, but it had an explosive impact.

In the painting, a thick-bodied timber rattlesnake has slithered up a tree trunk and reached the lip of a mockingbird nest. The coiled serpent, jaws unhinged, needle-sharp fangs bared, is poised to devour an entire family of four mockingbirds. The birds have just recognized the intruder. Their wings are raised in alarm, eyes wide, faces expressive. Are they too late?

Audubon's critics attacked him at once. Charles Waterton, a British naturalist, objected that rattlesnakes can't climb trees. George Ord, one of Wilson's closest friends, accused Audubon of exaggerating the size and angle of the snake's fangs to make the serpent more monstrous. This time he'd gone too far, they sniffed. The scene could not have occurred in nature, they scoffed. And they were wrong on both counts.

And can a rattlesnake climb a tree?
(Jerome Perez via ViralHog)

Audubon let his art speak for him, except to write a few lines in praise of mockingbirds:

Different species of snakes ascend to their nests, and generally suck the eggs or swallow the young; but on all such occasions, not only the pair to which the nest belongs, but many other Mocking-birds from the vicinity, fly to the spot, attack the reptiles, and, in some cases, are so fortunate as either to force them to retreat or deprive them of life.

And *Plate Number 21*, the portrait of a mockingbird family under siege, became one of the most famous bird paintings of all time.

*Three of the mockingbird specimens Darwin
collected from the Galápagos Islands
(Natural History Museum Images)*

CHAPTER SIX

Turning Point

*Charles Darwin and the Galápagos
Mockingbirds, 1835*

*My attention was first thoroughly aroused by comparing together
the numerous specimens . . . of the mocking-thrushes . . .*

—Charles Darwin, speaking of his development
of the theory of evolution

I n 1831, twenty-two-year-old Charles Darwin received a letter from
his Cambridge professor and mentor J. S. Henslow. It contained
the offer of a lifetime: Darwin was invited to serve as "ship's naturalist"
aboard the British ship HMS *Beagle*. Destination? Coastal areas of the
world. The crew would explore faraway places like Brazil, Tierra del
Fuego, South Africa, and New Zealand.

They would pay special attention to the coast of South America.
Great Britain was just developing trade relations with South America,
and accurate maps and charts were needed. Much of the coastline
remained unsurveyed.

Charles Darwin loved science, as did nearly everyone in his wealthy

British family. His uncle was a famous botanist, and his father, a doctor. Charles spent his boyhood roaming outdoors, exploring and studying the wonders of nature. His passion was geology—a science that deals with the history of the earth and its life as recorded in rocks. He daydreamed of witnessing volcanic eruptions and discovering new creatures. He was humble, brilliant, affable, inquisitive, and adventuresome.

But as he thought about the offer that had come his way, he was more than a little worried that two years at sea was too long. He had just graduated from college. Before that, he had dropped out of medical school, which had greatly displeased his father, and he was preparing for a career as a minister. Just when he had a future plan that silenced his father, along came a chance to see *the world*. His uncle Josiah tried to help, explaining to Charles's father that Charles was "a man of enlarged curiosity." How could he turn such an opportunity down?

He couldn't.

And he didn't.

Two days after Christmas 1831, seventy-four men squeezed onto a wooden ship with a deck barely bigger than a tennis court. They slept several to a room. Darwin shared his quarters with two other men. Aboard the HMS *Beagle*'s crew were cooks, rope masters, sailmakers, carpenters, surgeons, and instrument menders—and two parrots. As the naturalist, Darwin had the job of observing and collecting specimens of plants, animals, rocks, and fossils wherever the expedition went ashore. He was granted a very small cabin under the ship's forecastle for storing his specimens.

The *Beagle's* crew was young—almost everyone in his twenties or thirties—and extremely high-spirited. They had to delay departure for two days until drunken sailors could sober up from Christmas celebrations and stagger up the gangplank and onto the ship.

Among Darwin's first and most powerful discoveries was that the motion of the ocean waves made him seasick. He was often the first one off the boat and the last to reboard.

The crew visited Tierra del Fuego, South Africa, New Zealand, and the Azores but spent more than half their time mapping the coast of South America. Almost everywhere he went, Darwin encountered the Chilean mockingbird *Mimus thenca*. Tencas, as they were known, had been discovered and named by biologists in 1782. Perched atop scrubby plants, tencas filled the air with sound. Darwin admired their lyrical voices and amazing vocabulary. He wrote in his journal, "They are lively, inquisitive, active birds . . . possessing a song far superior to that of any other bird in the country."

The work took much longer than anyone had expected. In the summer of 1835, after more than four years of physical hardship and mental exertion, the *Beagle's* exhausted crew lifted anchor and turned westward for home. After traveling six hundred miles, they caught sight of their first stop: a cluster of volcanic mountains poking up through the seawater. They had reached the Galápagos Islands, described bleakly by Captain Robert FitzRoy, as "black, dismal-looking heaps of broken lava." But in the modern town of Puerto Baquerizo Moreno, they had reached a place where they could get off the ship and resupply before the final push home.

Like all the other crew members, Darwin was exhausted, but his curiosity drove him not to waste a minute. This would be his best chance to see an active volcano, and he felt sure that in a place so isolated there would be new plants and animals to collect. Although he did not get to see an active volcano, for the next five weeks Darwin hiked around four of the Galápagos Islands in the blazing sun, collecting specimens and returning to the *Beagle* only to sleep. He could barely believe what he was seeing. The plants and animals lived in, he wrote, "a little world within itself; the greater number of its inhabitants, both vegetable and animal, being found nowhere else."

Some creatures were unbelievably strange. There was a shark with a hammer-shaped head, lizards that could swim, and gull-like birds with bright blue feet. There was even a penguin that found some way to survive in the scorching heat of black volcanic rock. Darwin never knew what would be around the next bend in his path.

There were mockingbirds on each island, and they seized his attention. Darwin shot several specimens as he explored the San Cristóbal, Floreana, Isabela, and Santiago islands. To his surprise, he discovered that the specimens were different from island to island. On San Cristóbal, the mockers looked similar to the tencas he had seen on the coast of South America. But on the neighboring island, Floreana, the mockingbirds had darker breast markings, white bands on their wings, and longer beaks. He knew they were all mockingbirds by listening to the songs they sang, but beyond that, the differences aroused his curiosity the most.

Likewise, all the mockers on Floreana looked and behaved alike. But they were different than the mockers on San Cristóbal Island, all of which looked alike, too, but were likewise different from the mockers on the other islands. It seemed that each island Darwin visited had its own mockingbird.

Darwin was puzzled. How did it get that way? Had the mockingbirds somehow changed over time? At first it seemed impossible. Like most people of his time and religious upbringing, Darwin grew up believing that all the species of plants and animals on earth had not

Darwin Meets Giant Galápagos Tortoises

Darwin was astonished by one spectacular find: "As I was walking along I met two large tortoises, each of which must have weighed at least two hundred pounds: one was eating a piece of cactus, and as I approached, it stared at me and slowly stalked away; the other gave a deep hiss, and drew in its head. These huge reptiles, surrounded by the black lava, the leafless shrubs, and large cacti, seemed to my fancy like some antediluvian animals."

In fact, these giant tortoises were four feet long and weighed near five hundred pounds. Darwin took two of them back to England.

DARWIN TESTING THE SPEED OF AN ELEPHANT TORTOISE (GALAPAGOS ISLANDS).

(Alamy)

changed a bit since the moment they were created, some six thousand years before. They *couldn't* change. Darwin accepted this history; at least he did before his visit to the Galápagos. But a close inspection of all these new creatures raised disturbing questions that swam in Darwin's mind.

Back on board the *Beagle* in 1836, as the ship plied the equatorial Pacific waters for home, Darwin had a chance to compare the mockingbird specimens he had shot in South America and the Galápagos. What he saw widened his eyes and burst the bubble of his certainty. Darwin wrote, "In the Galapagos Archipelago, many even of the birds though so well-adapted for flying from island to island, are distinct on each; thus there are three closely-allied species of mocking-thrush [what Darwin called mockingbirds], each confined to its own island."

In other words, the mockers on each island were equipped with wings to visit the other islands, but they didn't make the trips. They stayed home. And over vast amounts of time, perhaps millions of years, they had somehow split into separate forms, or species.

What Is a Species?

A species is the most basic category into which living things are divided. A species of bird is a kind of bird—such as a northern mockingbird or, to use the scientific name, *Mimus polyglottos*. One well-accepted definition says that, for example, a group of birds is a species if members of the group can successfully reproduce only with each other. By this definition, if they tried to reproduce with birds that were not of their species, a young bird might be born, but it would not be able to produce young birds of its own.

The mockingbirds wrenched Darwin from the way he had been thinking. Now he reasoned that probably all mockingbirds in the Galápagos had descended from a single common ancestor because they were basically so similar to one another. He concluded that long ago maybe a single small colony of the tenca, the Chilean mockingbird, had flown or had been swept by powerful winds away from South America until they reached the Galápagos.

These mockingbird pioneers had given rise, by what Darwin called "branching descent," to three different species on three different islands in the Galápagos. He became convinced that the mockers had changed—or "evolved," as he put it—over time to suit their needs, until there was no point to jumping from island to island. This reasoning led Darwin to the ultimate conclusion that all organisms on earth had common ancestors, and that probably all life on earth had started with a single living creature.

There was still much to understand, especially how and why this had happened. But Darwin was convinced. As the *Beagle* reached Falmouth, England, on October 2, 1836, after nearly five years at sea, Darwin knew he was carrying ideas in the form of his mockingbird specimens that would be unpopular, because they were revolutionary. His most powerful idea was that new species arise naturally, by a constant process of evolution, rather than having been created—one time only and forever unchangeable—by God.

As the *Beagle* dropped anchor in England, Darwin and his mockingbirds were almost ready to share with the world one of the most controversial ideas of all time: the theory of evolution by natural selection.

Mockers of the Galápagos Archipelago Now

There are four mockingbird species on the Galápagos, one widely distributed, and the other three limited to one island each. All four are endemic to the archipelago, meaning you won't see them anywhere else on the planet. The four species differ from one another in plumage, eye color, and vocalizations. They never occur together on any single island.

The four mockingbirds are: (1) the Galápagos mocker, bold and inquisitive, which is often seen running along the ground; (2) the San Cristóbal mocker, the smallest of the four, which is endangered largely due to the work of cats and rats; (3) the Floreana mocker, plentiful in Darwin's day but now nearly extinct; and (4) the Hood mocker, aggressive and audacious, who demands food and water from tourists on Española Island and uses its long, curved beak to poke holes in seabird eggs for food.

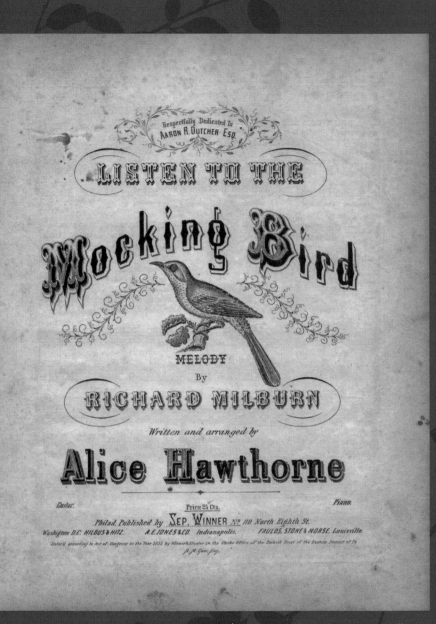

CHAPTER SEVEN

Listening to the Mockingbird

Septimus Winner and Richard Milburn, 1855

A real song . . . as sincere and sweet as the laughter of a little girl at play. —Abraham Lincoln

Septimus Winner was always on the lookout for song ideas. There was no telling when or how a melody or a lyric might pop into his head; he only knew a songwriter had to be ready because inspiration could strike anytime or anywhere.

Winner was a popular songwriter, music teacher, and music publisher in the mid to late nineteenth century in Philadelphia. He wrote many well-known songs in those years, usually under the pseudonym Alice Hawthorne—Hawthorne was his mother's maiden name, and she was related to the famous writer Nathaniel Hawthorne. Winner wasn't a wealthy man, but at the age of twenty-eight, he had his own studio in downtown Philadelphia, and his own music publishing company called Winner and Schuster. It could have been worse.

One afternoon while Winner was walking through the city, he came upon a crowd gathered on a street corner, alive with music and laughter. He worked his way up to the front and discovered the attraction was Richard Milburn, an African American musician who went by the name Whistling Dick. A barber by trade, Milburn was better known as a street musician. He strummed a guitar and whistled tunes on Philadelphia's streets while delighted onlookers tossed coins into his upturned hat.

The highlight of Milburn's performance was his mockingbird imitation. He wove an assortment of rapidly whistled musical phrases into a simple melody to which he always returned. It was a dead-on imitation of the lightning-fast trill of notes and shifting phrases performed by actual mockingbirds. Winner looked around the crowd. People were smiling and dancing and jabbing one another in the ribs, saying things like "Dick outmocked the mocker!"

As Milburn performed, Septimus Winner felt a tingle of inspiration. Why not arrange and publish a song out of Milburn's mockingbird melody, the part that he kept returning to? It would make a perfect chorus! He hustled back to his piano, pulled out a blank sheet of music paper, and jotted out a musical arrangement of the melody. Then he added lyrics to it. He titled the new tune "Listen to the Mocking Bird."

The lyrics told the tragic story of a grieving man whose lover, named Hally, has died. He mourns her loss by visiting her gravesite day after day and listening to a mockingbird sing, as they often did together when Hally was alive.

Winner published the song in 1855. In the years before sound recording equipment, publication of a song meant printing and selling copies of sheet music. The cover to the sheet music read "Melody by Richard Milburn" and "Arrangement by Alice Hawthorne." It is not known whether Milburn actually gave Winner permission to use the music, though his name was on the musical score.

The song sold slowly at first, and Winner gave up on it. A successful song could cross the country almost as fast as a news bulletin, and Winner knew what a hit looked like. This song wasn't it. He sold the copyright to another Philadelphia publisher for a trifling thirty-five dollars. The new publisher promptly erased Richard Milburn's name from the sheet and put out a new edition naming Septimus Winner as the only author.

Imitations

"Listen to the Mocking Bird" was first recorded in 1891, by Columbia recording artist John Yorke Atlee, who whistled the tune. Many versions have been recorded since by various artists, among them the Three Stooges, Chico Marx, Louis Armstrong, Fred Flintsone, Dolly Parton, and Barney the Dinosaur. It was Dolly Parton's father's favorite song. She has fond memories of singing "Mockingbird" with her dad.

Septimus Winner often complained about how difficult it was to make a living as a musician. If only he could have been just a little more patient, his financial worries might have disappeared forever. Inexplicably, the "mockingbird song" began to take off. More and more people could be heard humming and whistling Milburn's melody.

Before long, the sheet music was flying off music store shelves and burning up the mail routes. Listeners throughout the country were making up their own verses. People were wild about the song, especially in the South. For years afterward, Southern parents named their daughters Hally, after the tragic figure in the song.

It has never stopped selling. "Listen to the Mocking Bird" became one of the biggest hit songs in the history of music. Within the first fifty years, the sheet music sold a phenomenal twenty million copies in the United States and Europe. Abraham Lincoln praised it as "a real song . . . as sweet and sincere as the laughter of a little girl at play." King Edward VII of England said, "I whistled 'Listen to the Mocking Bird' when I was a little boy."

After Septimus Winner died in 1902, a Philadelphia newspaper wrote that he "had a capacity for seeing how to make music available for the masses that has probably never been approached by any man in the world."

And in the more than 160 years since it was published, the song that outmocked the mocker has never gone out of print.

(Birdchick.com)

CHAPTER EIGHT

A Visit to High Street Market

Philadelphia, Mid-19th Century

"Hope" is the thing with feathers—
That perches in the soul—
And sings the tune without the words—
And never stops—at all—

—Emily Dickinson

W hen Europeans immigrated to North America in the seventeenth century, they brought with them the practice of keeping birds. By the nineteenth and early twentieth centuries, birds had become the most popular indoor pets in the United States. According to writer Christal G. Pollock, "Like the televisions and radios of present day, birds brought beautiful song and a cheerful level of noise that was welcome in quiet, early American homes." The most common caged birds were brightly colored goldfinches, cardinals, and the greatest of singers, the mockingbird, which Philip Hone, mayor of New York City from 1826 to 1827, proclaimed as "the great leader of the feathered orchestra."

The caged bird trade was a booming business in the Philadelphia of the mid-nineteenth century. The *Philadelphia Business Directory* listed dozens of bird dealers, bird stuffers, and birdcage makers. Ornate birdcages were sold as works of fine art. On Wednesdays and Saturdays, families took outings to Philadelphia's clamorous High Street Market, one of the largest open-air markets in the world. The string of vendor stalls and storefronts stretched for almost a solid mile. Ducks and roosters called loudly from pens along the unpaved street. It was joyful bedlam!

Children dragged their parents past vendors and food stands and made their way to the caged birds. There they found cages stacked one on top of another. Inside, climbing the wires, were bright red cardinals and electric blue indigo buntings. Emerald green parrots used their powerful beaks and claws to hitch themselves along the wires of their cages.

Youngsters seemed to be magnetically attracted to the mockingbirds, who sang almost without stopping. Clustering around the cages, children tried to get the mockers to imitate them. They peppered the bird sellers with questions such as "Why does it keep singing?" and "How does it know all those songs?"

Many of the vendors knew very little about the birds they were offering for sale. They were products, nothing more or less. Typically, the vendor had paid a boy a pittance to dash across a lawn or field to snatch a baby bird from a mockingbird nest. Many of the nestlings were blind and naked except for a thin coating of gray down. The captured birds were placed into cages, fattened up for a few days, taught to imitate a few sounds, and taken to the bird stall at High Street Market. There, a singing adult mockingbird could fetch fifty dollars! That's what the vendor knew. He did his stumbling best to come up with answers while the kids kept hammering him with new questions.

Their curiosity was endless: "What do you feed it?" "Can you teach it a song?" "How many songs does it know?" And then, inevitably: "Please, Father . . . Mother . . . can we take it home? Please . . . Oh, thank you!"

How Do Mockingbirds Learn?

"Young mockingbirds start out babbling, making sounds, trying things out, much like human babies," says Dave Gammon, professor of biology at Elon University. "But if they're isolated in a room without language content, it will come out as babble. That's where neighboring birds come in, by providing songs to imitate. Mockingbirds are born with the *ability* to learn songs, but they need to learn the set list by imitating other birds." In other words, says Gammon, who has been studying mocker mimicry for fifteen years, "if they don't have the genetic wiring, with the brain set up as it is, they're not going to learn. But if they don't have the vocabulary, they're not going to learn either."

But there was a developing problem, one that Thomas Jefferson had recognized when he worried that mockingbirds would never reach Monticello. Records show that before the nineteenth century, free-roaming mockers were plentiful around Philadelphia. As word spread of the money to be made, so many mockingbirds were caged and sold that local wild populations were being decimated.

As the hit song "Listen to the Mocking Bird" wafted out of parlor windows throughout the United States and Europe, the real song was growing faint. By the late 1800s, there were only a few scattered recorded sightings in all of Pennsylvania, where the mocker was considered a rare breeding bird. The crisis continued into the new century, with only eleven nesting records for a five-county area in eastern Pennsylvania reported from 1901 to 1950.

Extinction—a lightly used word at the time—was becoming a real possibility, especially in the major cities of the eastern United States and in newly settled western cities such as Chicago and Saint Louis. There was too much money to be made by selling the nestlings. Mockingbird nests were too easy to reach, and the great feathered singers were almost too entertaining. In short, there was money to be made, and it was easy to make it. As Jefferson had foreseen, we were running out of mockingbirds.

Texas

USA 20c

Mockingbird & Bluebonnet

The mockingbird is the official state bird of five states, including Texas. (Shutterstock)

CHAPTER NINE

Parrott Shells and Minié Balls

The Mocker in the Civil War, 1861 to 1865

Civil War soldiers were often poorly trained, poorly led, and poorly armed, but no two armies ever went into the field equipped with better songs.

> --–Bobby Horton, *American Heritage* Magazine

Mockingbirds were favorite camp companions in the Civil War. Soldiers admired the trim gray warrior, who weighed little more than a bullet but whose clarion song lit up a camp. Mockers seemed to have a lot in common with soldiers. Fierce by nature, mockers tolerated no intrusion on their territory. They quickly attacked anyone or anything seen as a threat. Soldiers watched with delight as males competed for females through a ritualized dance, facing each other, squaring off, and rapidly bouncing sideways like boxers sparring, with heads held high, long tails twitching, and wings arched defiantly. *That's our bird*, the soldiers on both sides thought.

Civil War soldiers wrote in their journals and diaries of birds, thinking of them as food. Hungry soldiers caught and cooked the birds around their encampments and battlefields. One popular method of capture was to paint "birdlime," a glue-like material, on twigs and branches so that birds became stuck. Shooting at birds was forbidden both by Confederate and Union officers, not out of concern for birds, but to save ammunition. Penalties for disobedience could be severe— one diarist wrote of a soldier who, caught firing his revolver at a bird, was tied to a tree by his thumbs. Despite the risk, hungry soldiers took their chances and blasted away. However, mockingbirds were more frequently mentioned in diaries and journals as creatures whose song brightened the tedium and horror of war.

Brigadier General Alpheus S. Williams wrote to his daughter from a Union army camp in Tullahoma, Tennessee, on November 20, 1863: "We have very beautiful moonlights right now and an immensity of whippoorwills, and there are two mocking birds which begin their imitations every night in apple trees close to my tent. They mock everything from a frog to a crow. Some of their notes are beautifully sweet."

Major James A. Connolly, 123rd Illinois Infantry, wrote to his wife on June 9, 1864, from camp near Acworth, Georgia: "Brass bands are playing in every direction, and the mocking bird is making the leafy shade vocal with his attempts to imitate the brass music of 'Dixie,' 'Star Spangled Banner' &c &c."

Septimus Winner's "Listen to the Mocking Bird" was converted into a battle march used by both sides. Musicians hoisted their horns to their lips and charged alongside the soldiers into battle, keeping morale high. Music scholar Steven H. Cornelius notes that Union general Philip Kearney "often gave out whiskey to his band when they played ["Listen to the Mocking Bird"], which was frequent."

Drummers, many of them young boys, dictated the tempo of the action. Whether in camp or at home, on the march or in the field, everybody knew "Listen to the Mocking Bird."

The lyrics to the song were changed again and again to fit local situations. In 1863, the soldiers and residents of Vicksburg, Mississippi, found themselves hopelessly pinned down by Union forces who bombarded them with heavy Parrott shells. Gunboats lobbed more than 22,000 shells into the town. "Listen to the Mocking Bird" became the anthem of the Siege of Vicksburg.

The original lyrics to the song were:

I'm dreaming now of Hally,
Sweet Hally, sweet Hally;
I'm dreaming now of Hally,
For the thought of her is one that never dies . . .
Listen to the mocking bird,
Listen to the mocking bird . . ."

In Vicksburg, the lyrics became:

'Twas at the Siege of Vicksburg
Of Vicksburg, of Vicksburg;
'Twas at the siege of Vickburg,
When the Parrott shells were whistling through the air—
Listen to the Parrott shells,
Listen to the Parrott shells . . ."

And as hollow-based, elongated lead bullets called Minié balls came whistling by the thousands in their direction, and as food ran thin, still another verse to the mockingbird song marked their plight. It went:

Oh! well will we remember
Remember, remember
Tough mule meat, June sans November,
And the Minié balls that whistled through the air.
Listen to the Minié balls,
Listen to the Minié balls . . .

Since mockingbirds took no sides in the Civil War, both sides could proudly claim them as their own.

In the decades after the war, the bond between the mockingbird and Southern culture grew stronger. Five Southern states: Texas, Florida, Tennessee, Arkansas, and Mississippi named the mockingbird as their official state bird. Why did these states choose the image of the mockingbird to represent them on license plates, rest stops, and welcome signs? As the Texas proclamation says, ". . . [The mockingbird] is found in all parts of the state, in winter and in summer, in the city and in the country, on the prairie and in the woods and hills, and is a singer of distinctive type, a fighter for the protection of his home, falling, if need be, in its defense, like any true Texan."

*A fashionable woman of the early twentieth century wearing a hat
adorned with bird feathers (Library of Congress)*

CHAPTER TEN

Women to the Rescue

1865 to 1918

I know why the caged bird sings, ah me,
When his wing is bruised and his bosom sore,—
When he beats his bars and he would be free;
It is not a carol of joy or glee,
But a prayer that he sends from his heart's deep core,
But a plea, that upward to Heaven he flings—
I know why the caged bird sings!

—Paul Laurence Dunbar, "Sympathy"

At the close of the Civil War, there were no laws to protect birds from being shot for food or sport, or being collected and sold. And there was no real way to count how many individuals of a species were left, or to tell which species were safe and which needed help. When European whites had first arrived in North America, it was commonly supposed that there were so many birds in this paradise that it would be impossible to cause the complete extinction of a bird species. But some species were disappearing rapidly.

The best example is the passenger pigeon. On the morning of March 24, 1900, a fourteen-year-old Ohio farm boy named Press Clay Southworth was feeding cattle on his family's farm when he saw an unusual bird fly up into a tree and asked his mother's permission to use the family shotgun. When he presented her with the bird, she immediately recognized it as a passenger pigeon. The head and upper parts of the bird were bluish gray, and there were black streaks on its wings. The head and neck were small; the tail long and wedge-shaped. The wings, long and pointed, were powered by bulging breast muscles that could snap a bird through the air. It was a perfect engine for prolonged flight.

What Clay's mom didn't know was that the bird she held was potentially the last wild specimen of what might have been, barely a century before, the most abundant bird in the *world*. Bird artist Alexander Wilson estimated that one flock consisted of two billion birds; in 1813, John James Audubon watched a flock of pigeons, a mile wide and 240 miles long, pass over his head for three days. Before white settlement, more than a quarter of all the birds in what is now the United States were passenger pigeons. But they were tasty to eat and they destroyed crops by eating seeds. Farmers shot them and cast huge nets over fields to trap them by the thousands. Soon they were hard to find.

Caged passenger pigeons held on in zoos until September 1914, when Martha, a pigeon living in the Cincinnati Zoo and Botanical Garden, breathed her last. For several weeks, her worsening condition had been reported in local newspapers. In her last days she stared through the bars of her cage at a steady stream of visitors who had come to say goodbye not only to a specimen, but to a way of life. As essayist William Beebe wrote in 1906, "when the last individual of a race of living things breathes no more, another heaven and another earth must pass before such a one can be again."

More and more species were vanishing. Already, bird species like the Carolina parakeet and Ivory-billed woodpecker that were present at or near the battlefields during the Civil War were missing from those sites by the dawn of the twentieth century. Many other species were fading into memory.

The twentieth century began in a haze of smoke. Hunters gunned down bird species for profit and sport. A plumed-hat fashion craze seized women of the United States and Europe. Many American women wouldn't think of buying a hat without at least at least one long bird feather sewn into the band. Some women wore hats piled high with towering clouds of bird feathers. Seemingly overnight, there were fortunes to be made shooting down birds and selling the carcasses to hat-making companies.

The great egret and the snowy egret—wading birds who sported long white feathers—teetered on the brink of extinction. In 1903, an ounce of plumes—requiring the death of four birds—was worth twice as much as an ounce of gold. An ornithologist named Frank Chapman counted 700 hats, of which 542 had feathers sewn into them, during a two-block lunch walk in New York City. There were feathers from forty different species. More than five million birds were being massacred each year to feed the feathered hat industry.

Those who specialized in capturing and selling mockingbirds, however, found they could make more money keeping the birds alive and singing in cages than presenting them as feathered corpses atop a lady's head. Caged mockers could be seen and heard in parlors throughout the eastern United States. The birds were pleasant company, especially for women at home, often alone. Until recorded music and the radio, household life was much quieter. People kept caged birds in their

The aigrette—coiffed from an egret—
was the most popular of feathered hats.
(Courtesy of National Audubon Society)

kitchens or family living rooms, and they carried the cages with them from room to room throughout the day, hanging them at windows or in doorways.

The mocker was clearly in trouble, especially in and around eastern cities. The baby birds were too easy to collect. Their nests were typically just a few feet off the ground, and all one had to do was reach inside, pluck the babies out, and take them to market.

The author of one 1904 magazine article told of a shocking visit to a bird dealer in New York City. He wrote that "in one large cage [I] saw not less than sixty mockingbirds, some of them so young that when the cage was approached the poor birds hopped to the wire netting fluttering their wings and opening their mouths to be fed."

Mockingbird collectors traded tips: It was said that the best time to take a mockingbird from the nest was after a week, when the little birds were no longer blind and the parents had had a few days to feed them. Mockers needed dozens of spiders each day to keep up their protein. Kerosene in the cage removed lice. Iron from a rusty nail in their water cup could relieve diarrhea. Many collectors tired of cleaning the cages and feeding them daily, and tossed the birds outside, where they quickly perished.

Many mockingbirds were transported to Europe and sold in outdoor markets, especially the huge open-air market in Paris known as the Marché Saint-Germain. Bird lover Jules Michelet wrote of feeling sad and disgusted as he threaded his way along stalls heaped with small cages, stacked one on top of the next, in which birds barely had room to turn around. Especially touching were the mockingbirds, who mimicked other caged birds nearby. He called the mocker "an orchestra in himself."

With each decade of unregulated collection, the mockingbird's voice in the feathered choir grew fainter. Philadelphia's mockingbirds were dwindling fast by 1830, and the species had become rare in Delaware, New Jersey, and New York City by 1894. One farmer, convinced wrongly that mockingbirds were devouring his fruit crop, boasted of having killed 1,100 all by himself.

Ornithologist Henry Nehrling reported "great cargos" of young mockers entering the markets of Chicago and New York. Nehrling observed that while there was great demand for caged birds to provide singing entertainment for the humans who owned them, there were also "great losses and most died of mistreatment, neglect or ignorance."

While women's fashion drove the slaughter of millions of birds, women's activism would save them. The nation's first environmental protection groups formed in the 1890s to defend birds in what came to be called "the Plume War." In Boston, socialite cousins Harriet Lawrence Hemenway and Minna B. Hall hosted tea parties to inform their wealthy friends that birds were vanishing rapidly. Hall remembered, "We sent out circulars asking the women to join a society for the protection of birds." They called for a boycott of hats with feathers.

Portrait of bird champion Harriet Hemenway (Wikimedia)

Nine hundred women joined. Hemenway and Hall organized the "Massachusetts Audubon Society" to defend birds. The effort was named after the fabled bird painter John James Audubon—a title intended to inform the world that they were in the business of protecting birds. Audubon groups quickly formed in more than a dozen states. They were the first environmental protection groups in the United States.

The wealthy women used their social connections to pressure members of Congress and President Woodrow Wilson to pass laws to protect the songbirds that decorate our lives. They argued that the plume trade violated animal rights that applied to all living things. To cage a bird was cruel, unworthy of the cherished values of individual freedom and compassion for the weak.

And finally, like a ponderous door groaning on rusty hinges, the opposition gave way. In 1918, a bill protecting songbirds was passed by Congress and signed into law by President Wilson. The Migratory Bird Treaty Act made it a crime "to pursue, hunt, take, capture, kill, attempt to take, capture, or kill . . . any migratory bird" in the United States. Violations were punishable by fines up to five hundred dollars or up to six months in prison. Now, at last, with the dry scratch of a fountain pen on parchment, the great singers had a fresh chance to remain among the members of the earth's choir.

It was now a crime for a mockingbird "collector" to reach into a nest and snatch away eggs or defenseless nestlings. And even though mockingbirds did not migrate between separate breeding and wintering grounds, they were protected under the new law because they were not "game"—animals hunted for food and sport.

Cities emptied of mockingbirds were slow to recover their flocks. It would take a full century for the Philadelphia that had been trading in mockers, ever since Jefferson's day and before, to rebuild the population. And the Migratory Bird Treaty Act of 1918 would be challenged again and again in the years to come.

Mabel Osgood Wright: Gray Lady

One of the key figures in the Audubon movement to protect birds was Mabel Osgood Wright. Born to wealth, she was an early leader of bird protection efforts, especially in Connecticut. She wrote popular novels with bird protection themes and published stories for children.

In her 1907 book *Gray Lady and the Birds*, a boy has brought a cage containing an injured mockingbird into the kitchen of a Black cook at Gray Lady's Southern home. The cook informs the boy about the beautiful song and lively nature of mockingbirds. Gray Lady arrives, sizes up the situation, and says she will keep the bird until spring because it is injured. But she tells the boy and his friends that he must never confine wild birds in cages because it is against nature's way and causes unnecessary suffering.

But the women of the Audubon movement had organized a smart, tenacious defense. They had created a powerful new conservation tool that would still protect birds a century later. Energized by their passion, the women fought until it became a crime to kill hundreds of songbird species.

Including, in the nick of time, the northern mockingbird.

*Lullaby singer Lucy Cannady and her husband
(The English Folk Dance and Song Society)*

CHAPTER ELEVEN

Hush, Little Baby

1918 to 1937

Spiders and sowbugs and beetles and crickets,
Slugs from the roses and ticks from the thickets,
Grasshoppers, snails, and a quail's egg or two—
All to be regurgitated for you.

—Peter S. Beagle, *The Last Unicorn*

On Friday, August 23, 1918, British folklorist Cecil Sharp motored along a narrow two-lane Virginia highway, trying to look at the directions he'd been given while keeping his big car on the twisting mountain road ("I am sure nothing but a Ford could have done it," Sharp wrote). Sharp was traveling with his assistant, Maud Karpeles, through the Appalachian Mountains of North Carolina and Virginia to collect and record traditional songs—old mountain ballads, fiddle tunes, lullabies, and dance numbers—before they died out. Now he was driving along a ridgetop to visit Lucy Cannady of Endicott, Virginia, who had replied to his postcard with an invitation to visit.

When he stopped the car and got out, Lucy Cannady appeared at the doorway of her log cabin and looked over her visitors. Sharp was formally dressed in a suit and tie, a white cotton shirt, and a wide-brimmed fedora hat. He smoked a pipe. Lucy invited them inside. In an accent that sounded foreign to her, Mr. Sharp repeated the strange request he had made in his letter: Would she sing him the lullaby she sang to her baby each night? She agreed.

The lyric began with a bargain:

Hush! little Minnie and don't say a word,
Papa's gonna buy a mockingbird.
It can whistle and it can sing,
It can do most anything.

The song continued in rhyming couplets: "And if that mockingbird don't sing, Papa's gonna buy you a diamond ring." And if that diamond ring don't shine, Papa offers, in order, a looking glass, a billy goat, a cart and bull, a dog named Rover, and all manner of treats. The song's melody is so soothingly repetitive that Papa—or Mama, or any caregiver—can make up their own rhyming bribes until someone falls asleep.

As Sharp motored along the long ridges and shadowed valleys of the Blue Ridge Mountains, he discovered that many people seemed to know "The Mockingbird Song." It had traveled family by family, baby by baby, and nap by nap, through those mountains for hundreds of miles over the course of many years. Sharp was amazed by the sheer quantity of songs these people knew. A few weeks after his visit with Lucy Cannady, Sharp met forty-nine-year-old Julie Boone in Micaville, North Carolina. She taught him twenty-nine tunes he had never heard before, including a new version of "The Mockingbird

Charlie and Inez Foxx's "Mockingbird"

In 1963, brother and sister Charlie and Inez Foxx recorded their own version of "Mockingbird." It is rooted in the famous nursery rhyme collected by Cecil Sharp decades before, but the Foxxes didn't sing to a baby: They sang to each other. In an interview, Inez later recalled how she converted the lullaby into a rhythm and blues number: "I said to [Charlie], 'You should do something else to try to pick up the beat of the song . . . Let's try this: Hey, everybody, have you heard?' I said, 'Now you repeat it with me but don't sing it with me.' He said, 'What you talking about?'" The Foxxes' version has a fierce, pounding beat and clever lyrics. Their recording shot up the US popularity charts, peaking at Billboard's number 7 on September 7, 1963. Many artists covered the song. The Foxxes' dreams of stardom came true overnight. Twenty-one tours followed, including stints with the Beatles and the Rolling Stones. "I don't know if it was God's will or what," Inez said later, "but I just picked it right up."

(Getty)

Song." The words were slightly different, leading with "Hush up, baby, don't say a word, Papa's going to buy you a mockingbird."

Nineteen years later, in March 1937, an African American school-teacher named Annie Brewer from Montgomery, Alabama, sang her version to folklorist John Lomax, who slid a microphone in front of her and recorded her voice into his machine.

Annie Brewer's version, probably because it was the first to be recorded, was the "Hush, Little Baby" that became known around the world. Lomax's recording of the song was made publicly available through the US Library of Congress.

It was also included in a book of American folk songs titled *Our Singing Country*. But like Richard Milburn before her, Annie Brewer received no income for her contribution to the song, though the Library of Congress did ascribe the song's authorship to her.

In the 1950s and 1960s many famous musicians, including Pete Seeger, Nina Simone, Joan Baez, and Peter, Paul and Mary, recorded the song. Though it sold millions of copies, no member of Annie Brewer's family received payment for singing her mockingbird song into John Lomax's recording machine. As researcher and musician Harvey Reid, put it, "Annie Brewer's contribution, though essential and undoubtedly the source of all the others, has been lost, ignored, forgotten or removed."

Cecil Sharp heard Lucy Cannady's lullaby about buying a mocking-bird in 1918, the same year the Migratory Bird Treaty Act was passed. That made it the last year that any papa or mama or anyone else could legally buy or sell a mockingbird. By then it would have been hard to find a mocker for sale anyway, for they were becoming scarce in eastern marketplaces. The great feathered mimic had been all but wiped out in the eastern portion of its range after more than 150 years of punishing collection by those who caged them and sold them for a song.

There was a final irony in the mockingbird lullaby. After all the sleepless nights that mockingbirds had caused humans over the centuries with their loud, all-night singing, here at last was a bit of compensation—a mockingbird tune that could put a baby to sleep.

*Amelia Laskey in the field with a duck she had just banded
(Courtesy of the Tennessee State Library and Archives)*

CHAPTER TWELVE

Amelia Laskey, Citizen Scientist

1920s to 1970s

The mockingbird's invention is limitless;
he strews newness about as casually as a god.

—Annie Dillard, *Pilgrim at Tinker Creek*

O n the morning of Tuesday, August 1, 1939, a silver-haired woman carefully lifted a nine-day-old male mockingbird out of a nest in a Nashville, Tennessee, park and took him home to study. She set him up in a cage wrapped in wire mesh and rolled the cage out onto her screened-in porch so the bird could see, hear, and speak. She took notes from a chair just outside the cage. She named him Honey Child, which soon became abbreviated to H. C.

Amelia Rudolph Laskey, born in 1885, was a nationally known and highly respected bird scientist, especially admired for her beautifully researched life histories of bird species. She married and moved to Tennessee with her husband, Fredrick, in 1921. There, Amelia planted a garden on their four acres of land, described by her friend Katherine

Goodpasture as "a natural home for wild things that grew riotously . . . but with an indescribably delicate beauty." The garden attracted a multitude of birds of many species. Her curiosity about these birds blossomed into a passion. She joined the Tennessee Ornithological Society in 1928 and obtained a federal bander's permit so she could track the movements of birds.

Laskey was a prime example of the "citizen scientist," an expert who did not study her specialty in a classroom, but whose curiosity drove her to find ways to increase our knowledge.

Mockingbirds were Laskey's favorites. She listened carefully to the mockers in her garden and throughout her neighborhood, wondering, as had many before her, how mockingbirds get their songs. Are they born with the songs they sing or do they learn songs throughout their lives? Do they really mimic the songs of other bird species? If so, why?

These were time-worn questions, often debated, and up to this point, never resolved one way or another. Ornithologist J. Paul Visscher wrote in 1928, "it seems probable that a mockingbird does not as a rule consciously mimic songs, but only possesses an unusually large series of melodies which it calls forth in wonderful perfection." Dr. George R. Mayfield wrote in 1934 that "the mockingbird inherits his repertory from many generations back." Dr. Witmer Stone, writing in 1935, threw another log on the debate's fire by confessing that he had "always thought that many so-called imitations recorded in print are really not imitations at all."

Amelia Laskey believed that the best way to inform the debate was to carefully observe and listen to the songs of a single mockingbird for a long period, recording the bird's song selection through detailed notes. She knew she was the perfect investigator: tireless, passionate, sharp-eyed. She was financially able to work from her home. Most of all, she was equipped with the patience to silently observe and record a single mockingbird's song for days and months on end. Now all she needed was a bird to study.

Enter H. C.

The young mockingbird that Amelia Laskey lifted from the nest remained all but silent for his first three weeks with the Laskeys. He lived in a cage 42 inches long, 14 inches wide, and 19 inches high. The cage was mounted on a wheeled table and rolled out onto a screened-in porch so that H. C. could see and hear other birds. The Laskeys opened H. C.'s cage door and allowed him to hop and fly freely throughout the house at least once a day. The young mocker had a terrific appetite, opening his beak whenever he wanted to be fed and gobbling down portions of pabulum, ground beef, peeled apples, and raisins. The Laskeys did not try to make a pet of H. C., though they found him very entertaining. Sometimes the vacuum cleaner or washing machine or Mr. Laskey's whistling would touch off a vocal response from H. C.

On August 19, 1939, H. C. delivered his first feeble musical notes. The next day he was silent, but the day after that, "he sang softly with closed beak for ten minutes."

Citizen Science

The first half of the twentieth century brought years of fresh, excited interest in studying birds. Bird conservation groups sprouted overnight. Binoculars magnified birds by several times and could be bought by mail order for seven dollars. Thousands of children joined Junior Audubon clubs, learning to draw and paint birds, but only if they signed a pledge that said "I promise not to harm our birds or their eggs, and to protect our birds whenever I am able." In 1934, artist Roger Tory Peterson's book *A Field Guide to the Birds* was published, containing paintings of several hundred bird species, with tips on how to identify them. The book fit snugly into the back pocket of a pair of trousers.

Eight years earlier, in 1926, Cornell University hired Arthur Allen as the country's first ornithology professor, and he immediately started a noon-time bird identification program for local women who wanted to know more about the birds that visited their gardens and feeders.

[L] Arthur Allen's lunchtime birding seminar (Courtesy of the Cornell Lab of Ornithology); [R] (Alamy)

Day by day H. C.'s singing grew in strength and confidence. On October 29, the three-month-old mocker sang on and off between 7:20 PM and 10 PM. Then, for the next two nights, H. C. sang all night long by the light of a full moon. Amelia stayed up with him, scribbling notes in the dark.

On December 11, H. C., now aged four and a half months, sang the first songs that reminded Laskey of other bird species. She recognized fragments of songs from the downy woodpecker, Carolina wren, blue jay, catbird, flicker, cardinal, starling, bobwhite, and canary. Sometimes his singing was raucous; other times barely a whisper. Laskey noted, "Songs included whistles, trills, warblings, and squawks."

With the onset of spring 1940, H. C.'s songs grew louder and lasted longer. Laskey's April 12 notes proudly announce, "A varied and indescribable performance; songs are loud and long, starting at 5:30 AM (CST), continuing all day." On another unforgettable June day, H. C. burst forth with 143 ecstatic songs in sixteen minutes, challenging Amelia's hand speed as she took notes.

In 1944, after more than four years of studying H. C., Amelia Laskey compiled her notes into an article titled "A Mockingbird Acquires His Song Repertory." It was published in *The Auk*, a prestigious magazine about bird studies. "Listening to many Mockingbirds," she wrote, "year after year, one learns there are portions of the song that are 'true Mockingbird' song because they are peculiar to that species and are given by all in that part of their range."

Experiments on free-flying mockingbirds, combined with the data she developed at home in her study of H. C., suggested that mockers began life with an inherited set of tunes that remained constant

The first page of Laskey's notes describing how Honey Child developed his songs (Warner Park Nature Center, Metro Nashville Parks and Recreation)

no matter where the birds went or to whom they listened. In other words, there seemed to be a core set list of mockingbird hits.

But there was also evidence that mockers were truly padding their repertoire by mimicking other species. Laskey reported, "My notes contain a number of records for several species which were answered immediately [by H. C.] in the same call notes or song each had just uttered."

She was particularly impressed by two incidents that seemed to show a mockingbird's ability to mimic with a purpose. On February 27, 1942, a flicker—a species of woodpecker that often finds its food on or near the ground—flew to the Laskeys' driveway and prepared to drink from a puddle in the snow. H. C. was watching intently from his cage. As the woodpecker bowed to drink, H. C. gave a perfect imitation of the flicker's unmistakable *wicka!* call. Two days later it happened again; a flicker landed in a tree near H. C.'s cage. H. C. immediately delivered several rounds of *wicka!* How did he know the flicker's call? As Laskey noted, "at both demonstrations the Flickers were silent"—so he couldn't have picked up the call from the flicker in question—"and the mockingbird had not been singing." Her conclusion: "H. C. is very observant of life about him."

Her answer to the nature v. nurture question—Are mockingbirds born with their songs or do they learn them by imitating others?—seemed to Laskey to be a bit of both.

Do Mockingbirds Keep Learning Songs Throughout Their Lives?

Dave Gammon, professor of biology at Elon University, says he owes a great debt to Amelia Laskey. Her observations, clearly documented, are still important nearly a century later. "Everyone and their dog had avoided mockingbird songs, but Ms. Laskey had the patience to develop data that is still important today," Gammon says.

Gammon's work has shown that, like human babies, mockingbirds learn most of their songs early in their lives. Gammon did a long-term study of mockingbird songs and found that in their fourth year they were singing pretty much the same songs they had been singing in year one. "There's a window for learning that narrows after about a year," says Gammon. "They can continue to learn throughout their lives, but it's easier at the beginning."

Gammon says that mockingbird songs change depending on where you hear them. "A North Carolina mockingbird will not sound the same as a California mockingbird," says Gammon. "And the birds they imitate will be different, too. I started out in Texas and heard mockingbirds imitate great-tailed grackles, but never common grackles. Then I moved to North Carolina, and I hear them imitate common grackles all the time."

H. C. died in 1954 after having lived with the Laskeys for fifteen years and four months—about twice as long as mockingbirds typically live in the wild. The Laskeys had treated him well. They proudly witnessed each year's molt and enjoyed the gush of H. C.'s springtime sounds.

Amelia Laskey continued to study mockingbirds throughout her long life. Her nationally acclaimed observations drew upon data from more than 250 nests, from over 900 banded nestlings, and from observations she made of individual birds. Along the way, she discovered that mockingbirds hold two types of territory: summer and winter. The summer breeding territory contains the nest and is defended fiercely by the male while the female lays and incubates eggs. The winter territory is usually smaller than the summer grounds, often centered around a food source such as a fruit-bearing tree or bush.

Laskey's work as a citizen scientist didn't stop with mockingbirds—she developed creative experiments to inform big questions about many bird species. She bought a telescope, bundled herself up, and joined a "moon-watching" program, counting birds in migration as they passed in front of the moon. When television towers were first erected, and many birds smashed into them, Laskey became a pioneer in watching for casualties at TV towers and developing recommendations to protect birds.

As her friend Katherine Goodpasture wrote of her, "This lady lived with quiet modesty and scientific integrity. . . . She shared her vast experience in natural history gladly and was ever generous with her means."

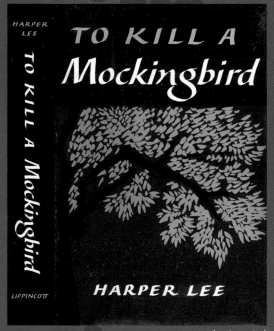

(Wikimedia)

CHAPTER THIRTEEN

To Kill a Mockingbird

1960

Shoot all the blue jays you want . . . but remember it's a sin to kill a mockingbird.

—Atticus Finch in *To Kill a Mockingbird* by Harper Lee

Many people know about mockingbirds only from the title of a book. The first novel by author Harper Lee, titled *To Kill a Mockingbird*, was published without fanfare in 1960. It was about racial injustice in a small Alabama town. Few had heard of the author. But Lee's book became an instant classic, thundering up the bestseller charts, remaining at number one for eighty weeks and winning the Pulitzer Prize for Literature in 1961. The plot and characters were based on Ms. Lee's observations of family and neighbors in her hometown of Monroeville, Alabama, in 1936, when she was ten years old.

The book's title drew attention to mockingbirds. Why did Harper Lee choose a mockingbird? Why not *To Kill a Robin* or *To Kill a Cardinal*? What was so special about mockingbirds?

In one pivotal scene from *To Kill a Mockingbird*, lawyer Atticus

Finch gives his young children, Scout and Jem, air rifles as holiday gifts. With the gift comes an ominous warning. "I'd rather you shot at tin cans in the back yard, but I know you'll go after birds," Atticus says. "Shoot all the blue jays you want, if you can hit 'em, but remember it's a sin to kill a mockingbird."

Scout was deeply disturbed by her father's warning:

That was the only time I ever heard Atticus say it was a sin to do something, and I asked Miss Maudie [their neighbor] about it.

"Your father's right," she said. "Mockingbirds don't do one thing but make music for us to enjoy. They don't eat up people's gardens, don't nest in corncribs, they don't do one thing but sing their hearts out for us. That's why it's a sin to kill a mockingbird."

In other words, we humans have a special ability to enjoy a good singer, and the northern mockingbird has a special ability to sing a good song. To kill a mockingbird would be to destroy a duet.

With his famous warning, Atticus Finch, a lawyer, was advising his children that it was all right to break the law selectively. Since the passage of the Migratory Bird Treaty Act of 1918, it had been illegal "to pursue, hunt, take, capture, kill, attempt to take, capture or kill . . . any migratory bird" in the United States. Both mockingbirds and blue jays enjoy the same protection under this law. Getting caught breaking this law is punishable by stiff fines and even imprisonment. Yet, as Allegheny College English professor John MacNeill Miller asks pointedly, "If the law doesn't discriminate between blue jays and mockingbirds, why does Atticus Finch?"

The northern mockingbird was already the official bird of five states when Ms. Lee's book was published, but the story brightened

the spotlight on the great singer. People and mockingbirds seem to share
a unique bond in the novel. It is the strength of that partnership that
makes mockers too special to kill. More than a murder, it would be a
sin, a spiritually ruinous act. Mockingbirds symbolize innocence and
beauty. Atticus and Miss Maudie tell Scout and Jem that it's a sin to kill
a mockingbird because these birds cause no harm to human life—they
just sing. Killing a mockingbird would be an act of senseless cruelty.
Several human characters in the novel can be seen as innocent mockingbirds, especially Tom Robinson and Boo Radley. They are fragile, kind,
and moral individuals who are misunderstood and often mistreated by
their prejudiced society—and, in Tom's case, ultimately destroyed by it.

The story has inspired some readers to want to know more about
mockingbirds. They want to hear for themselves what is so special
about the famed song. Harper Lee's *To Kill a Mockingbird* is, above all,
thought-provoking. It is one of the most thoroughly discussed books
of all time. To date, it has sold more than forty million copies and has
been translated into forty languages. And, for many, it puts the mocker
on the map.

(Alamy)

CHAPTER FOURTEEN

Improvising

Mockers on the Move, 20th and 21st Centuries

All through the night the cries of curlews and plovers and knots, of sandpipers and turnstones and yellowlegs, drifted down from the sky. The mockingbirds who lived on the island listened to the cries. The next day they would have many new notes in their rippling, chuckling songs to charm their mates and delight themselves.

—Rachel Carson, *Under the Sea-Wind*

I think of the mockingbird as our native jazz musician. . . . No other area of the world, as far as I know, has a bird so devoted to improvisation.

—Earl Vickers, American audio engineer and sound programmer

O ne of the grandest construction projects in human history was the building of a network of interstate highways throughout the United States in the 1950s and 1960s. This colossal enterprise—crisscrossing the nation with 46,000 miles of concrete in all—sliced up the landscape and paved great quantities of bird habitat.

The most adaptable species found ways to get by. Swallows nested under highway culverts and bridges; sparrows and robins found food in manicured grass; songbirds perched and sang from guardrails.

Perhaps the biggest winner of all was the northern mockingbird. Mockers scavenged food and nesting material from the multiflora rosebushes that were planted along superhighway divider strips to reduce glare. An average multiflora plant produces a million seeds a year, as well as gobs of tasty berries. Many species of birds began to gravitate toward this multiflora, but mockingbird parents were among the most common and best known for constructing their nests in the thick tangle of roots and leaves. They followed the highways, steadily expanding their breeding range. Tom Fegely, a newspaper reporter in Lehigh, Pennsylvania, wrote in 1986, "As a youngster . . . I don't recall seeing any mockers in the Southern Lehigh area in the 1950s. . . . Now there's not a community, housing development, country town, or suburban dooryard that doesn't have one or more mockingbird families in residence."

The mockingbird population has been shifting northward over the past century or so, as the earth's warming climate has turned areas that were previously ice covered and barren into possible places for a mockingbird to winter. It's all about food: New plants bearing fruits and berries have taken root in warmer climates, and insects now tunnel through and bore into places that were, until recently, too cold for them to survive aboveground in the winter.

The Mockingbird and Jazz Music

Among New Orleans jazz musicians there is a special signal that tells players it's time to pick up their instruments and play. "It's like: C'mon, let's muster, let's get to the bandstand and be ready to do it," says guitarist Matt Bell. The signal is a sequence of four notes, commonly blown on a horn or whistle. But there is another species, also found throughout New Orleans, that knows this signal: the northern mockingbird.

Peter Yaukey, a bird expert and geography professor at the University of New Orleans, told National Public Radio that musicians may have learned the signal from mockers. "That particular whistle sounds very much like the phrases a mockingbird would use," he said. "It's repetitive. It's got the right pitch. It's got the right amount of substance to the notes."

No matter who came up with the four-note signal, it has been passed from musician to musician and bird to bird ever since. And it adds up, for jazz musicians and mockingbirds have much in common. Both stay up late playing songs of romance, and both are known for improvising. Both experiment with sounds. On a spring day, New Orleans neighborhoods are alive with the sound of mockingbirds, and saxophones, spinning their love songs into the sultry air.

(*Alamy*)

Mockingbirds have spread to northern states, such as Michigan, Minnesota, Wisconsin, Maine, and even to southern Canada, where winter bird counts in Quebec and Ontario show population increases.

The first mockers reached frosty Maine in the 1890s. Tags wrapped around the legs of museum specimens suggest that these first birds arrived in cages. Early museum records inform us that the birds "escaped." It's hard to know what that means. Perhaps the owners, annoyed by a bird's song or weary of cleaning cages, simply tossed them out into the snow. Or perhaps the birds really did escape, darting out through a briefly open door or window. One specimen was shot. It took another half century for Maine mockingbirds to become permanent residents.

Mockers are becoming city birds, or, as scientists say, "urban adapted." We continue to create mockingbird habitats when we clear forests, replacing trees with buildings, gardens, hedges, lawns, and of course, highways. We trim a hedge; they build a nest in it. Rush hour too noisy? They put off their singing until their songs can be heard above traffic. Bright city lights allow mockers to feed their nestlings after dark. A telephone pole becomes a soundstage for a male who hasn't yet found a mate.

Another plus is that mockers are prolific breeders. Most mate for life. It's not unusual for a couple to bring out three broods—batches of eggs—during a breeding season. One amazing female mockingbird laid twenty-seven eggs in a single springtime. Author and Texas naturalist Roy Bedichek expressed his admiration for mockers' parental

Mockers as DJs: Constantly Sampling

The mockingbird's song continues to fascinate scientists. In May 2021, a team of four scientists published an article in the journal *Frontiers in Science*, describing the rules behind the mocker's song. They focused on the way the singing bird moves from one cluster of syllables to the next. They found that the mocker uses musical strategies like those of human musicians. "When you listen for a while to a mockingbird, you can hear that the bird isn't just randomly stringing together the melodies it imitates," said neuroscientist Dr. Tina Roeske. She explained that "it seems to sequence similar snippets of melody according to consistent rules." The scientists identified four strategies that mockers use in switching from one sound group to the next. They are changing pitch, changing timbre, stretching the transition, and squeezing the transition. "We have so much in common with mockingbirds," said Dr. David Rothenberg in an interview. "Both species learn from what we hear around us. We both cut and paste music together. We're like disc jockeys, constantly sampling."

devotion. He wrote of a mockingbird couple who attempted to build a nest in his backyard during a drought, calling them "indomitable." They worked "like beavers" to rear young successfully when parents of other species had given up.

Some mockers migrate during winter months, but usually not for long distances. The trend seems to be either to stay put and try to survive the winter—a gamble—or to migrate short distances and try to stay just ahead of blizzards and cold snaps. The mockers who stay put and survive gain a head start on those who leave, since they're already occupying the breeding grounds when spring arrives. Also, they have saved the energy required for a migratory flight. But if they miscalculate and become trapped inside a bitter winter storm, it could mean their lives. Nature operates on a very thin margin.

A University of Florida student flinches under the attack of an irate parent mockingbird.
(Photo by Gustavo Adolfo Londoño Guerrero)

CHAPTER FIFTEEN

"They Know Me!"

The Genius of Mockingbirds, 2007

We think . . . our experiments reveal [mockingbirds'] underlying ability to be incredibly perceptive of everything around them and to respond appropriately when the stakes are high.

—Professor Doug Levey, the University of Florida

The campus of the University of Florida (UF) in Gainesville, Florida, is one of the best places anywhere to study northern mockingbirds. They sing from parking lots and lawns, from classroom buildings, and from trees that line the open spaces of the campus.

So it is hardly by accident that there are apt to be several mockingbird studies taking place at UF at any given moment.

In the spring of 2007, Christine Stracey, then a PhD student at UF, led a huge study aimed at learning more about how and why mockingbirds seemed to be doing better in urban areas—cities and suburbs—than in farms and fields. The study included more than one thousand mocker nests scattered throughout three residential neighborhoods, two parking lots, two pastures, and two wildlife preserves. Stracey was in charge of monitoring all the neighborhoods. "I got to know the home-owners really well," she recalls, "especially one lady who was always out-side gardening when I showed up. She was so excited about our study."

A mockingbird nest had been woven at about eye level in the limbs of a holly tree that overhung the lady's sidewalk. Inside the nest were four newborn mockingbirds, orange mouths gaping open and necks stretched upward, aching for the next meal to arrive. Stracey's job was to remove each little bird from the nest, weigh it for the study, and place it back in the nest. But day by day, it got harder to do the work because whenever she pulled up and got out of her car, the par-ent mockingbirds spotted her and went berserk, scolding her harshly and dive-bombing her head to drive her away from their nestlings. On the other hand, they left the gar-dening lady completely alone. She witnessed the attack from her garden and invited Stracey indoors for a breather one day when the ordeal was through.

University of Florida PhD student Christine Stracey insisted that the mockingbirds within her study group could distinguish her from other humans. (Photo by Gustavo Adolfo Londoño Guerrero)

"That's amazing," the woman said. "Those birds recognize you."

It was true, and it was becom-ing an ordeal. With each visit, the mockers seemed to identify Stracey earlier and attack her sooner.

After one heart-pounding visit to the holly tree, Stracey told her colleague Dr. Doug

Levey about the difficulties she was having collecting data. The problem, she said, was the birds knew her. Levey answered firmly, "No way. Those birds don't *know* you," to which Stracey replied, "Yes way. They do." To her, it was clear as day. They attacked her personally. They even attacked her car when she got close to the tree. They knew her, and they felt threatened by her.

By the end of their conversation, Doug Levey started to take Stracey's complaint seriously. If it could be proven that mockingbirds could recognize individual predators, Levey knew the finding would be important to science. Levey had studied cognition—brain activities—in birds. He knew many of the most famous studies on bird intelligence had focused on crows and jays, birds who had learned to perform tasks with human tools. But what was learning, when it came to birds? It wasn't necessarily about using human tools. Learning was about adapting to get the things that you needed.

If the UF team could show that mockingbirds could discriminate among predators, that would put mockers on the elite bird intelligence map. It would also explain why mockingbirds attacked some intruders and not others. They seemed to be thinking, *Why should I waste energy defending nests against creatures that pose no threat?*

The UF team designed a simple experiment to test whether mockingbirds could recognize specific humans. It required very little funding and took only a month. A group of biologists and volunteers divided into five teams of two, one approaching the nest and the other videotaping the action. The walker advanced slowly and steadily, reaching the nest in fifteen seconds. Then the walker would reach up and place a hand on the lip of a mockingbird nest and hold it there for fifteen seconds. After they'd done this for four days, the two researchers switched roles, with the ex-videographer now becoming the walker, approaching the nest slowly and steadily—and just standing beside the nest for fifteen seconds before walking calmly away.

The result? UF's biologists found that the mockers recognized the danger posed by the nest-touching human sooner each day; they met and attacked the increasingly familiar intruder at greater distances from the nest. And they barely reacted to the second human—the researcher who had not yet touched the nest.

"We found that after a single trial, the mockingbirds had learned which humans were a threat"—the people touching the nest—"and which ones weren't," said Dr. Scott Robinson, another team leader for this experiment.

They repeated the experiment again and again. With each trial, the birds seemed to get smarter and more confident. Eventually, the mockers could pick out the humans who had touched their nest from a crowd of a hundred people. The researchers tried to fool the mockers; they changed the clothes they were wearing and altered their hairstyles. Some days they wore hats; others they didn't. But they couldn't fool the birds. To a parental mockingbird, only one thing mattered: the safety of the nestlings. Time after time, the mockers attacked only those humans who had committed the sin of touching their nests.

Shortly after the UF team published their experiments in the journal *Proceedings of the National Academy of Sciences*, British news source *The Guardian* recognized the importance of the Florida experiments: "The extraordinary behavior . . . is thought to be the first published account of wild animals in their natural setting recognizing individuals of another species."

The University of Florida experiments furnished the scientific data to prove what Native Americans and other bird observers such as Jefferson, Audubon, and Laskey had noticed for centuries: Mockingbirds seem to have a sixth sense that gives them, among other things, a jump on predators.

Dr. Scott Robinson summed it up as follows: "We of course took this as one more line of evidence that birds are way, way more intelligent than we think. They have very small brains, but those brains are really powerful. . . . We think one of the reasons why [mockingbirds] are so common, they're so successful, they do so well in urban environments, is they can learn very exactly who is a threat and who isn't, even within predators. And in fact, this is one of the reasons why we think that being called a birdbrain is not an insult at all."

Epilogue

So it would seem that mockingbirds are thriving. They have everything going for them: Superior intelligence. A brilliant communication network. A varied diet. A knack for adapting to changing circumstances. High egg production. And who could ask for a better parent than a mockingbird? These birds defend their young fiercely, often driving away invaders many times their size. "They will soon have a cat batting at the air and spinning around in circles until it becomes a blurry whirling dervish of hissing fur and flying claws," wrote Florida reporter Budd Titlow. On top of that, they are protected by law against human mischief, and new habitat is becoming available to them as the earth warms.

And yet, the overall mocker population is dwindling. It's not a crisis yet—many songbirds are in deeper trouble than mockingbirds—but the loss is worth noticing. As they expand their range in the warming North, their numbers are steadily trickling down in the South. The northern mockingbird is assigned a Continental Concern Score of only 8—20 being the highest alert—by Partners in Flight, which tracks and measures bird endangerment. That means there is no emergency, and yet according to the North American Breeding Bird

Survey, "northern mockingbird populations declined by about 21% from 1966 to 2015."

Why the drop-off? Some experts point to changes in the way we farm these days. As tractors, combines, and other farm machinery become monstrously large, fields have become larger too. "Huge, flat open land is not good for mockers," says Christine Stracey. "They need fences for singing and nesting."

Songbirds throughout the world are struggling to adapt to a warming climate. One study showed that birds are laying eggs at an average rate of six days earlier per decade. Birds are migrating earlier in the spring, arriving earlier at their breeding grounds, and departing later. Some birds in Europe are skipping migration all together.

One day in October 2015, Nevada nature writer Michael Branch stepped onto his front porch to fill his lungs with sweet desert air, when "an unmistakable flash of white-barred wings appeared in the sagebrush." Then there was a song, then another, and another—loud, brassy, and confident. Branch likened the performance to "an avian karaoke machine."

It was the first mockingbird Branch had seen in more than ten years of living in the desert. He described his guest vividly: "It struts around on its tall, skinny legs, proudly holding its long tail up high behind it, acting like it owns the place. It perches on the lawn furniture and appears bothered when we head outside." Branch knew the bird was on the move, probably pausing only to snatch up insects that scuttle through the sage and to pluck berries from fall fruit bushes.

Branch was grateful for the moments they shared. When the

Mockingjays: Feathered Spies

For those who read for pleasure and adventure, the northern mockingbird will continue to be a messenger in song well into the future. In 2010, author Suzanne Collins completed her wildly successful Hunger Games trilogy. The final volume is titled *Mockingjay*, after a bird specially equipped for a vital task.

The mockingjay was a bird created when jabberjays (made up by the author for the trilogy) mated with mockingbirds (real birds—our mockers). The offspring were called mockingjays.

Jabberjays were bred by the brutal government of Panem to spy on enemies and rebels of the Capitol, and could memorize and repeat entire human conversations. But when the rebels found out that their conversations were being intercepted, they sent back messages riddled with false information. The government found out and angrily abandoned the birds to die in the wild. There, male jabberjays bred with female mockingbirds, producing the mockingjay. The new breed was able to repeat both human melodies and birdsong. The mockingjay became a symbol for the rebel cause.

(Alamy)

mockingbird was gone, he remembered it as "a messenger between seasons and between worlds."

Messenger through song: That sounds about right. It was the same role Mockingbird played for the Zuni people when passing out languages to those emerging from under the ground and into the blinding sunlight.

Through the centuries the northern mocker has inspired artists and poets, storytellers, musicians, and explorers. Soothed our children to sleep with a lullaby, kept us awake all night with a bottomless set list, and finally, just when we've dropped off to sleep, jolted us back awake for just one more tune.

The mocker has fired our imaginations and triggered our consciences. It has made us better than we were. For thousands of years our cocky gray partner has shared with us a bargain—sustenance for a song.

Messengers through song: That's us—a duet. Paired beings who can learn the songs of other species—not just our own—and make use of them.

And so we make our case. Who is the most important North American bird species in the lives of humans?

Our duet partner. The northern mockingbird.

WHAT YOU CAN DO TO
HELP MOCKINGBIRDS AND
OTHER SONGBIRDS

The northern mockingbird is not an officially listed endangered species, at least not yet. So it might seem a bit rushed to devise a plan to save a species that seems to be doing well, especially when there are other birds in worse shape.

But the mocker population is shrinking, even as they expand their range. And why wait until it's too late to act, especially for a bird that has meant so much to our history?

I asked several biologists and conservationists the following question: "What can we do to help mockingbirds?"

Here are a few suggestions:

- If you have a yard, seed it with native plants and shrubs with berry-laden branches. This is especially important in winter, at a time when mockingbirds eat mostly berries. Plant it with grasses that allow mockers to forage on the ground. Seed in grasses with blades that will not get too high for the mockers to see over it.

Cats, engineered to stalk and pounce, kill millions of birds each year in the United States. (iStockPhoto)

- Bring your cats indoors. Admittedly, this is hard for families whose cats have been raised outdoors. But those outdoor feral and pet cats, engineered to stalk and pounce, kill 2.6 billion birds each year in the United States and Canada, even when well fed. A worldwide study in which cats were fitted with video cameras showed that a cat kills about four birds a month. If you must let your cats go outside, avoid the early morning and later in the day during the breeding season when birds are most active, especially younger birds that are not accomplished fliers. Put a bell on your cat.

- Avoid pesticides and herbicides.

- Defend—and expand—bird protection laws. These laws have been attacked many times throughout the years. In August 2020, federal judge Valerie Caproni upheld the 1918 Migratory Bird Treaty Act, scolding those from the US

Department of the Interior who sought to weaken the law. "It is not only a sin to kill a mockingbird," she lectured. "It is also a crime. . . . That has been the letter of the law for the past century."

- Finally, learn all you can about mockingbirds. Learn their songs. See if you can pick out the songs of other birds embedded in the mockers' melodies. Try to learn the songs of the species they are imitating.

These birds are endlessly fascinating. They know an amazing number of songs. As naturalist and photographer Diane Cooledge Porter says, "I think if you try listening to the mockingbird, its song will become dear to you. And someday if you move out of the mockingbird's range, you will miss it."

ACKNOWLEDGMENTS

First, I thank my parents, Darwin Hoose and Catherine Williams. In 1952, when I was five, my parents brought home a parakeet. We named it Tweety. It was a blue budgerigar with a wickedly hooked beak. Tweety was a born showman. Dad started each day shaving, whistling as he scraped the stubble away from his chin. Tweety matched him note for note. So I confess: A caged bird was the first bird I ever loved.

I also offer thanks to Cuban biologist Carlos Peña for perspective as to which birds Columbus must have heard—and certainly didn't hear—in Cuba.

I appreciate Linda Pearsall, Joe Little, and Michael Lee Bierly for pointing me to information about the life and times of Amelia Laskey. Thanks to Sandi Bevins for locating Ms. Laskey's field notes and providing photographs.

Thanks to Dr. Herb Wilson, who provided information about the historical distribution and movement of northern mockingbirds in Maine. Bill Hancock of Maine Audubon generously shared data from the Maine Christmas Bird Count. Many thanks!

Thanks to Keith Ouchley, who introduced me to the controversy of the rattlesnake who reached the lip of a mockingbird's nest in Audubon's *Plate 21*. Thanks likewise to Kelby Ouchley for expert perspectives on the impact of warfare on birds during the Civil War.

In a series of telephone interviews, Christine Stracey shared her story of the groundbreaking University of Florida experiments showing that mockingbirds can recognize specific individual humans. Dr. Dave Gannon helped me understand how and why mockingbirds learn songs. And, in a telephone interview, David Rothenberg shared exciting new perspectives about musical mockingbird strategies for transitioning from one cluster of notes to the next.

Janice Stillman, editor of *The Old Farmer's Almanac* at Yankee Publishing, encouraged me to write about mockingbirds by publishing my mocker profile in the *Almanac*.

Richard Machlin, researcher extraordinaire, helped me piece together an outdoor market scene in 1855 Philadelphia.

Matt Bell taught me guitar from his home in the Lower 9th.

Hannah Hoose and Ruby Hoose, the finest of daughters, listened supportively to countless mockingbird soliloquies.

Kirsten Cappy, my partner in young peoples' literature for several decades, came through once again with two sparkling and insightful reads.

Literary agent Jennie Dunham carefully and artfully found a home for this book. No bump in the road seemed to exhaust her creativity. Her comments on the manuscript made this a much better book.

My editor, Melissa Warten, was the rock that I returned to again and again in this project. A lesser editor might have been discouraged by, among other factors, COVID-19. But not Melissa. She kept us on course. And she is a miraculous photo finder. Thank you!

Sandi Ste. George, my wife, listened to my reading of every chapter and commented as only a great reader can. Sandi fell deeply in love with the mockers who sing near our New Orleans apartment. There is no pleasure greater than to walk with my duet partner through the Lower Garden District as mockers fill the springtime air with song.

SOURCE NOTES

The COVID-19 pandemic made traveling for face-to-face interviewing difficult, so I conducted most interviews by phone.

There is a wealth of material in books and magazines about mockingbirds. Mockingbirds have inspired storytellers, poets, artists, and musicians for thousands of years. The birds are not shy. They have been written about in thousands of sources. The challenge to an author writing about the connection between mockers and humans is deciding what to leave in and what to omit. Where's Eminem? Where's Walt Whitman?

Here are a few selected sources of general information about "the King of Song."

SELECTED BOOKS

Ackerman, Jennifer. *The Genius of Birds.* New York: Penguin Press, 2016.

Bahr, Donald. *How Mockingbirds Are: O'odham Ritual Orations.* Albany: State University of New York Press, 2011.

Doughty, Robin W. *The Mockingbird.* Austin: University of Texas Press, 1988.

Sibley, David Allen. *What It's Like to Be a Bird: From Flying to Nesting, Eating to Singing—What Birds Are Doing, and Why.* New York: Alfred A. Knopf, 2020.

INTRODUCTION: WHAT'S THAT SOUND?

The street scene in which a mockingbird fends off an attack by a colossal predator happens frequently during the spring breeding season. Nature photographers pad their portfolios with images of mockingbirds heroically driving off hawks, owls, even eagles in defense of nests and fledglings. I have observed groups of tourists throughout New Orleans oblivious to the mockingbirds singing rapturously within a few feet of their group. This book is to help humans be as observant as mockingbirds.

A description of the syrinx and how it functions is found at Ackerman pages 142–144.

CHAPTER ONE: FOUR HUNDRED TONGUES

The stories come from the website Native Languages of the Americas (native-languages.org), a Minnesota nonprofit corporation dedicated to the preservation and promotion of endangered American Indian languages.

"How Mockingbird Became the Best Singer" is a Mayan legend about how mockingbirds learned to sing. The entire legend can be found at FirstPeople.us.

Find Jovita González's story "Mocking Bird" in *The Woman Who Lost Her Soul and Other Stories: Collected Tales and Short Stories*. Published by Arte Público Press, 2001. It's available online at Scribd.com.

CHAPTER TWO: SETTLERS, EXPLORERS, AND MOCK-BIRDS

Columbus's insistence that nightingales could be found in the New World: See Ana Torfs, "The Parrot & the Nightingale, a Phantasmagoria" (anatorfs.com/projects/The-Parrot-the-Nightingale -a-Phantasmagoria).

Catesby and Lawson: "Lawson and Catesby—First Explorers" by Suzannah Smith-Miles, Special to the *Moultrie News*, May 4, 2017.

Mark Catesby's Legacy: Natural History Then and Now by M. J. Brush and Alan H. Brush. Published by the Catesby Commemorative Trust, 2018.

CHAPTER THREE: "DICK SINGS"

The chapter header quote is the first of three stanzas to the poem "Poem 254" by Emily Dickinson, ca. 1861, from *The Complete Poems of Emily Dickinson*, edited by Thomas H. Johnson. Published by Little, Brown and Company, 1955. Nineteen US Presidents have had pets: "Jefferson and his Mockingbird" by Brian Fischer for Presidential Pet Museum (presidentialpetmuseum.com/blog/jefferson-and-his-mockingbird/).

Captivity sidebar: "Jefferson's Ornithology Reconsidered" by Matthew R. Halley, Drexel University, from a presentation given October 2015 at the American Philosophical Society Museum.

Jefferson's mockingbirds' names and behavior, especially Dick: monticello.org/site/research-and-collections/mockingbirds.

Margaret Bayard Smith's description of Dick and Jefferson together is found in *The First Forty Years of Washington Society* by Margaret Bayard Smith. Published by Charles Scribner's Sons, 1906.

CHAPTER FOUR: MOCKER V. NIGHTINGALE— FULL-THROATED RIVALRY

Thomas Jefferson's goading letter to Abigail Adams is found at the Massachusetts Historical Society, Adams Family Correspondence, Volume 6, Thomas Jefferson letter to Abigail Adams, Paris, June 21, 1785.

Nightingale v. Mockingbird. A fascinating look at "the growing need to establish a national identity in nineteenth-century

American literature" is found in Gabe Cameron's master's thesis, *The Establishment and Development of the Mockingbird as the Nightingale's "American Rival"* (dc.etsu.edu/cgi/viewcontent.cgi?article =4677&context=etd).

"To an English Nightingale" by Maurice Thompson in *Poems, pp. 10–18*. Published by Houghton-Mifflin, 1892.

CHAPTER FIVE: A RATTLESNAKE CLIMBS A TREE

There are several fine biographies of Audubon. My favorite is *Under a Wild Sky: John James Audubon and the Making of* The Birds of America by William Souder. Published by Milkweed Editions, 2014.

Alexander Wilson described his life and brilliant work in his American *Ornithology*. For more information about Wilson's life and times, see *Alexander Wilson: Naturalist and Pioneer, a Biography* by Robert Cantwell. Published by Lippincott, 1961.

Read Audubon's detailed description of the northern mockingbird that accompanies *Plate 21* in *Birds of America* at audubon.org/birds-of -america/mocking-bird.

To learn more about Audubon's work—especially his love of shooting birds to gain specimens and his positioning of specimens on screens to create dramatic poses—see "The Dark Side of Audubon's Era, and His Work" by Anne Raver in the *New York Times*, March 30, 1997.

To see a video of a rattlesnake hunting from a treetop, search "rattle-snake in a tree" online—the resulting photos and videos are striking!

CHAPTER SIX: TURNING POINT

Charles Darwin's findings remain controversial after nearly two hundred years, sparking tension between religious and scientific communities. What was he like to know and work with? Author Deborah

Heligman turns to diaries and letters to find clues to the man behind the theory. See *Charles and Emma: The Darwins' Leap of Faith*. Published by Henry Holt and Co. Books for Young Readers, 2009.

To learn more about the central role that mockingbirds played in Darwin's theory of evolution, read "To Understand a Mockingbird: Specimens That Sparked Darwin's Theory of Evolution" by Ian Sample, in the *Guardian*, November 14, 2008 (theguardian.com /science/2008/nov/14/evolution-charles-darwin).

CHAPTER SEVEN: LISTENING TO THE MOCKINGBIRD

Septimus Winner was famous enough to merit a biography: *The Mocking Bird: The Life and Diary of Its Author, Septimus Winner* by Charles Eugene Claghorn. Published by Magee Press, 1937.

The story of Winner hearing Whistling Dick Milburn imitating a mockingbird is also well-known. But there is still research to do. Did Milburn permit Winner to turn his street performance into a published sheet of music? Did Milburn agree to Winner's sale of the song to another publisher for a paltry sum? And what was it about the song that made it an overnight sensation?

CHAPTER EIGHT: A VISIT TO HIGH STREET MARKET

Information about bird-keeping habits of early European immigrants to North America can be found in the article "Companion Birds in Early America" by Christal G. Pollock, in *Journal of Avian Medicine and Surgery*, volume 27, number 2, 2013, pp. 148–151.

I consulted *Gopsill's Business Directory* and *The Philadelphia Business Directory*—both circa 1870—to find lists of bird dealers, bird stuffers, cage makers, and cage sellers. My friend Richard Machlin provided research that helped me reconstruct Philadelphia's High Street Market.

Much information about the mockingbird's role in Civil War camp life derives from Kelby Ouchley's *Flora and Fauna of the Civil War: An Environmental Reference Guide*, published by Louisiana State University Press, 2010. The author wrote of the wreckage done by warfare to plants, animals, and ecosystems. The diary entries are fascinating and heartbreaking.

For alternative lyrics to "Listen to the Mocking Bird," read the article of that title by Ted Widmer in the *New York Times*, November 5, 2013 (opinionator.blogs.nytimes.com/2013/11/05/listen-to-the -mockingbird/).

Every state's official bird is listed at statesymbolsusa.org/categories /bird.

CHAPTER TEN: WOMEN TO THE RESCUE

The chapter header quote is from Paul Laurence Dunbar's poem "Sympathy," from *The Complete Poems of Paul Laurence Dunbar*. Published by Dodd, Mead and Company, 1913. The "Myth of Superabundance," as it came to be called, is the erroneous belief that the earth has more than sufficient natural resources to satisfy humanity's needs, and that no matter how much of these resources humanity uses, the planet will continuously replenish the supply.

The story of Clay Press Southworth shooting the last uncaged passenger pigeon is told in detail by author Christopher Cokinos in his book *Hope Is the Thing with Feathers: A Personal Chronicle of Vanished Birds*, published by Warner Books, 2000.

The Plume War. After watching their peers go from having bird feathers in their hats to wearing whole dead birds, concerned women founded the Massachusetts Audubon Society in 1896. Read all about it in *She's Wearing a Dead Bird on Her Head!* by

Kathryn Lasky. Published by Little Brown Books, 1997. For more advanced readers, see *The Audubon Ark: A History of the National Audubon Society* by Frank Graham Jr. Published by Alfred A. Knopf, 1990.

Gray Lady and the Birds: Stories of the Bird Year for Home and School by Mabel Osgood Wright. Published by the Macmillan Company, 1907.

CHAPTER ELEVEN: HUSH, LITTLE BABY

The chapter header quote is from *The Last Unicorn* by Peter S. Beagle. Published by New American Library, 1968. English folklorist Cecil Sharp made several song-collecting trips to the United States. He took a special interest in songs of the Appalachian Mountains. He wrote almost daily in his diary during his American trips. These entries paint a rich portrait of the people that he and his traveling colleague, Maud Karpeles, met. The entries can be found at cecilsharpinappalachia.org/1918travels.html.

You can hear Charlie and Inez Foxx perform their version of "Mockingbird" by visiting youtube.com/watch?v=g47_NI1CWNQ.

For a thought-provoking biography of the song "Hush, Little Baby," see Harvey Reid's blog "The Troubadour Chronicles," February 2020 (woodpecker.com/blogs/hush_little_baby.html). Reid is a scholar and musician who has sharp, well-researched opinions about folk music and many other topics.

CHAPTER TWELVE: AMELIA LASKEY, CITIZEN SCIENTIST

The chapter header quote is from *Pilgrim at Tinker Creek* by Annie Dillard. Published by Harper's Magazine Press, 1974. Amelia Laskey described her experiments with the caged mockingbird H. C. in a lengthy article appearing in the magazine *The Auk*, Volume 61, April

1944, pp. 211–219. The article is titled "A Mockingbird Acquires His Song Repertory."

Katherine Goodpasture's detailed and admiring obituary of Laskey appears in *The Auk*, Volume 92, April 1975, pp. 252–259.

Interviews with Joe Little and Michael Bierly informed me about Amelia Laskey's contributions to the Nashville birding scene in her day.

For more information about citizen scientists, consult the sidebars scattered in chapters three and four of my book *The Race to Save the Lord God Bird*, published by Macmillan, 2014.

CHAPTER THIRTEEN: TO KILL A MOCKINGBIRD

The chapter header quote, alongside the famous passages in which Atticus Finch gives his children air rifles and a mockingbird warning, and in which Scout asks Miss Maudie for an explanation, appear in chapter 10 of *To Kill a Mockingbird* by Harper Lee. Published by Lippincott, 1960 (and in many editions since).

CHAPTER FOURTEEN: IMPROVISING

Information about mockingbirds' use of the plant multiflora rose is at "In Praise of Multiflora Rose" by Les Line, in *Audubon* magazine, June 14, 2008 (audubon.org/news/in-praise-multiflora-rose).

Merlin is a free app designed by the Cornell Lab of Ornithology that helps identify birds in four ways. It's simple to use. If you have a photo of a bird you don't know, tap "Photo ID" to identify your bird. If you hear a bird that you don't recognize or can't see, tap "Sound ID" to record the bird you are hearing, and Merlin shows you which species you are most likely hearing, in real time! Or you can pick "Explore Birds," and Merlin will show you all the birds most likely to be around you on that day.

Much of this chapter was informed by three recorded interviews with Dr. Christine Stracey, a central figure in the 2009 University of Florida experiments. Dr. Stracey is now an assistant professor of biology at Guilford College in Utah.

The scientific report of the University of Florida experiments by Douglas J. Levey, Christine M. Stracey, and Scott K. Robinson, among other researchers, showing that mockingbirds recognize individual humans, was published as "Urban Mockingbirds Quickly Learn to Identify Individual Humans" in *Proceedings of the National Academy of Sciences*, Vol. 106, No. 22, June 2, 2009, pp. 8959–8962.

A good article about the experiments is found in the *Guardian*. The article, written by science correspondent Ian Sample, is titled "Mockingbirds Bear a Grudge Against Particular People." It was published in the May 18, 2009, issue and can be found at theguardian.com /science/2009/may/18/mockingbirds-human-recognition.

WHAT YOU CAN DO TO HELP MOCKINGBIRDS AND OTHER SONGBIRDS

You can read about Judge Valerie Caproni's decision to uphold the Migratory Bird Treaty Act in an August 12, 2020, *Washington Post* article titled "Quoting 'To Kill a Mockingbird,' Judge Strikes Down Trump Administration Rollback of Historic Law Protecting Birds." See www .washingtonpost.com/climate-environment/2020/08/11/quoting -kill-mockingbird-judge-struck-down-trumps-rollback-historic-law -protecting-birds/.

INDEX

A

Adams, Abigail, 34
adaptation
    to cold temperatures, 7
    to human life, 7, 8
    to superhighways, 104
    to urban areas, 7, 106–7
aggression
    toward bird predators, 4, 41
    when defending nests, 23, 41,
       110, 112–14
Allen, Arthur, 92
Alsop, Fred, 35
Angelou, Maya, 101
Appalachian Mountains, 83–84
Arkansas, 72
Armstrong, Louis, 59
Atlee, John Yorke, 59
Audubon, John James
    artistic reputation, 42–43
    bird species discovered by, 43
    controversy over paintings,
       43–45

    encounter with Wilson, 40–41
    finishes *Birds of America*, 43
    organization named after, 80
    painting style, 41
    passenger pigeon counts, 76
    passion for painting birds, 40
    Plate Number 21, 38, 44–45
    praises mockingbird's song, 34
Audubon, Lucy, 39
Audubon groups, 80, 92
The Auk, 93

B

Baez, Joan, 86
bald eagle, 8
Barney the Dinosaur, 59
battle marches, 70–71
The Beatles, 85
Beebe, William, 76
Bell, Matt, 105
berries, 22, 104, 123
Beverley, Robert, 19

binoculars, 92

bird conservation groups, 92

bird intelligence studies, 113–14

birdlime, 70

bird painters. *See also* Audubon,
  John James; Wilson, Alexander
    in the 1800s, 39–45
    Mark Catesby, 21, *21*

bird protection laws, 125

birdshot, 40

bird species
    casualties from TV towers, 96
    declining populations, 6,
      75–77
    defined, 52
    discovery of, 43
    environmental stresses, 6–7
    hunted for hat industry, 77
    hunted for sport, 77
    Laskey's research on, 96
    legal protection for, 80–81
    loss of habitat, 6

The Birds of America (Audubon),
  43

Boone, Julie, 84–86

Boston, 79–80

Branch, Michael, 118–20

branching descent, theory of, 53

breeding behaviors, 23, 107

breeding territory, 96

Brewer, Annie, 86

British exploration of the New
  World, 19

C

caged birds
    as Civil Rights image, 101
    in early 19th century, 42
    in early 20th century, 77–79
    in late 1700s, 25–26
    in mid-19th century, 63–66
    at the White House, 25–30

Canada, 7, 106

Cannady, Lucy, *82*, 83–84

Caproni, Valerie, 124–25

Caribbean islands, 17

Carolina parakeet, 77

Catesby, Mark, 20, 21, *21*

cats, 124

Cencontlatolly, 11–12

Chapman, Frank, 77

Cherokees, 11–12

Chicago, 79

Chilean mockingbird, 49–53

Cincinnati Zoo and Botanical
  Garden, 76

citizen scientists, 90, 92

Civil Rights Movement, 101

Civil War soldiers, 69–72

Cleveland, Grover, 27

climate change, 6, 104, 118

cognition in birds, 113

Collins, Suzanne, 119

Columbus, Christopher, 17–19, 34

Connolly, James A., 70

Continental Concern Score, 117

Coolidge, Calvin, 27

Cornelius, Steven H., 70

Cornell University, 92

D

Darwin, Charles

    discovers giant tortoises, 51

    early love of science, 47–48

    Galápagos Island discoveries,
      49–52

    HMS *Beagle* voyage, 48–52

    theory of evolution, 53

Delaware, 79

Dick (White House mockingbird),
  25, 29–30

diet, 22, 104, 123

distribution and habitats

    breeding territory, 96

    expansion of, 6–7

    northern climates, 7, 104–6

    range map, 6, 6–7

    summer breeding territories,
      96

    superhighways, 104

urban areas, 7, 106–7

    winter territories, 7, 96, 108

Duboule, Denis, 6

E

Edward VII, King of England, 60

eggs, 23, 23, 107

environmental protection groups,
  79–80

evolution, theory of, 53

extinction, 66, 75–77

F

Fegely, Tom, 104

female mockingbirds

    broods per season, 107

    eggs laid by, 23, 23, 107

    parenting skills, 23

    songs, 5

Ferdinand (King of Spain), 17

A Field Guide to the Birds
  (Peterson), 92

Finch, Atticus, 99–100

flicker, 93

Flintstone, Fred, 59

Floreana mocker, 50, 54

Florida, 72

folios, 40

Foxx, Charlie and Inez, 85,
  85

Franklin, Benjamin, 8
fruits and berries, 22, 104, 123

G

Galápagos Islands
    exploration of, 49–52
    giant tortoises on, 51
    mockingbird species on, *46*,
        50–53, 54
Galápagos mocker, 54
Gammon, Dave, 95
geology, 48
giant tortoises, 51
Glover, Thomas, 20
González, Jovita, 14
Goodpasture, Katherine, 89–90, 96
Gray Lady and the Birds (Wright),
    81
great egret, 77
gunpowder, 39–40

H

Hall, Minna B., 79–80
hats, feathered, *74*, 77, 78, 79–80
hawks, 4, 41
Hawthorne, Alice, 57
Hayes, Rutherford B., 27
Hemenway, Harriet Lawrence,
    79–80, *80*
Henslow, J.S., 47

herbicides, 125
High Street Market, 64
highways, 102–4
Hispaniola (the Dominican
    Republic and Haiti), 17
HMS *Beagle* voyage, 47–52
Hone, Philip, 63
Honey Child (H.C.) (Laskey's
    mockingbird), 89, 91–95
Hood mocker, 54
Hopi folklore, 12, 13–14
Hunger Games trilogy, 119
"Hush, Little Baby," 86

I

I Know Why the Caged Bird Sings
    (Angelou), 101
insects, 22
interstate highways, 102–4
Isabela Island, 50
Isabella (Queen of Spain), 17, 34
Ivory-billed woodpecker, 77

J

Jackson, Andrew, 27
jazz music, 105
Jefferson, Martha, 28
Jefferson, Thomas
    admiration for mockingbirds,
        25, 34

mockingbirds owned by,
25–26, 28–30, 40
moves back to Monticello, 30
moves to the White House,
26–28
views on slavery, 29
Juana (Cuba), 17
Junior Audubon clubs, 92

K

Kearney, Philip, 70
King Edward VII, 60
King Ferdinand, 17

L

language development, myths
about, 11–14
Laskey, Amelia, 88
article published in *The Auk*,
93
background, 89
joins moon-watching program,
96
recommendations for bird
protection, 96
studies on mockingbird songs,
90–96
Lawson, John, 20
Lee, Harper, 99–101
legends, Native American, 11–15

Levey, Doug, 112–13
Library of Congress, 86
Lincoln, Abraham, 27, 60
Lincoln, Tad, 27
Linnaeus, Carl, 20
"Listen to the Mocking Bird," 56
becomes anthem of Siege of
Vicksburg, 71–72
converted into a battle march,
70–71
original tune, 58
popularity of, 59–60
subsequent versions, 59
Lomax, John, 86

M

Maine, 106
male mockingbirds, 5, 23
Marché Saint-Germain, 79
Maricopa folklore, 12
Martha (passenger pigeon), 76
Marx, Chico, 59
Massachusetts Audubon Society,
80
mating behaviors, 5, 107
Mayan culture, 12–13
Mayfield, George R., 90
Michelet, Jules, 79
Michigan, 106
migration, 7, 96, 108

Migratory Bird Treaty Act of 1918, 80–81, 100, 124–25

Milburn, Richard, 58, 59

Miller, John MacNeill, 100

Mimidae family of mimic thrushes, 4

Mimus polyglottos, 5

Mimus thenca, 49

Minié balls, 71

Minnesota, 106

Mississippi, 72

"mock-bird," 20

mocker-human partnership
    with Civil War soldiers, 69–72
    food for song, 8, 22, 120
    in *To Kill a Mockingbird*, 101
    with New World colonists, 19–20
    with southeastern US settlers, 21–22

"mockers" nickname, 5

"A Mockingbird Acquires His Song Repertory" (Laskey), 93

mockingbirds. *See also* northern mockingbird
    Chilean (tencas), 49–53
    on Floreana Island, 50
    on Galápagos Islands, 46, 50–54
    on San Cristóbal Island, 50
    scientific name, 4–5

"The Mockingbird Song"
    first recorded version, 86
    known in Appalachian Mountains, 84
    rhythm and blues version, 85
    versions of, 84–86

Mockingjay (Collins), 119

Monticello, 26, 30

"moon-watching" program, 96

multiflora rosebushes, 104

myths, Native American, 11–15

N

Native Americans
    mockingbird myths and legends, 11–15
    musical instruments and songs, 22

natural selection, 53

Nehrling, Henry, 79

Neruda, Pablo, 9

nestlings
    captured and sold, 42, 64, 78
    feeding of, 23, *23*
    legal protection for, 80

nests
    building, 23

defending, 23, 41, *110*, 112–14

eggs, 23, *23*, 107

A New Voyage to Carolina (Lawson), 20

New Jersey, 79

New Orleans jazz, 105

New World

colonists, 19–20

exploration of, 11, 17–19

New York City, 79

nightingale, 35

confused with mockingbird, 18

description of, 33

role in European literature, 33–34

transatlantic rivalry with mockingbird, 33–36

North Carolina, 20

northern mockingbird

acrobatic skills, 22, 35

adaptability, 7, 8, 104, 106–7

aggressive behaviors, 4, 23, 41, *110*, 112–14

bond with humans (*See* mocker-human partnership)

breeding behaviors, 23, 107

as caged pets, 25–26, 42, 63–66, 77–79

collectors of, 78

confused with nightingale, 18

Continental Concern Score, 117

courage of, 22, 41, *42*

depicted in *Plate Number 21*, *38*, 44–45

diet, 22, 104, 123

evolution of, 7

female, 5, 23, *23*, 107

habitat (*See* distribution and habitats)

how to help, 123–25

intelligence, 111–15

legal protection for, 80- 81

male, 5, 23

mating behaviors, 5, 107

"mock-bird" name for, 20

"mockers" name for, 5

named official state bird, 72

in Native American myths and legends, 11–15

parenting skills, 107–8

physical appearance, 2, 14–15, *15*, 19, 23, 32, 34, *106*, 118

population declines, 65–66, 79, 86–87, 117–18

population increases, 6–7, 81

portrayed in *Mockingjay*, 119

portrayed in *To Kill a Mockingbird*, 100–101
predators, 4, 41, 113–14
range map, 6, 6–7
rivalry with nightingale, 33–36
songs (*See* songs and calls)
as symbol of innocence and beauty, 101
University of Florida studies on, 111–15
white stripes on, 14–15, *15*

O

"Ode to Bird-Watching," 9
O'odham folklore, 12
Ord, George, 44
Our Singing Country, 86

P

Parisian outdoor markets, 79
Parrott shells, 71
Partners in Flight, 117
Parton, Dolly, 59
passenger pigeon, 76
Peña, Carlos, 18
pesticides, 6, 125
Peter, Paul and Mary, 86
Peterson, Roger Tory, 92
Philadelphia
caged bird trade, 63–66

mockingbird population, 65–66, 79, 81
Plate Number 21, 38, 44–45
plumed hats, 74, 77, 78, 79–80
"The Plume War," 79–80
poison ivy berries, 22
Pollock, Christal G., 63
Porter, Diane Cooledge, 125
predators
ability to distinguish among, 113–14
hawks, 4, 41
Pueblo folklore, 12

Q

Queen Isabella, 17, 34

R

Radley, Boo, 101
Randolph, Thomas, 26
rattlesnakes, 44, *44*
red-tailed hawks, 4
Reid, Harvey, 86
RitzRoy, Robert, 49
Robinson, Scott, 114, 115
Robinson, Tom, 101
Roeske, Tina, 107
The Rolling Stones, 85
Rothenberg, David, 107

S

San Cristóbal mocker, 50, 54

Santiago Island, 50

Seeger, Pete, 86

Sharp, Cecil, 83–87

Shasta Indian mythology, 12

Sibley, David, 8

Siege of Vicksburg, 71–72

Simone, Nina, 86

slavery, 29

Smith, Margaret Bayard, 29

snowy egret, 77

songs and calls

    attracting females with, 5

    debates over, 90, 94

    descriptions of, 3, 34, 70, 93

    duration of, 5, 34, 93

    genetic wiring for, 65

    geographic variations, 95

    how young birds learn, 65

    imitation of non-bird sounds, 5

    imitation of other birds, 5, 65, 90, 93–94

    inherited tunes, 93–94

    link with jazz music, 105

    in Native American myths and legends, 11–15

    Richard Milburn's imitation of, 58

    rules behind, 107

    studies on, 65, 90–96, 107

    syringeal muscles used for, 5–6

    training birds in, 25–26

South America, 47, 49

South Carolina, 20

Southworth, Press Clay, 76

Spanish explorers, 11, 17–19

species, defined, 52

Stone, Witmer, 90

Stracey, Christine, 111–13, *112*, 118

subscriptions for paintings, 40, 42–43

syringeal muscles, 5–6

syrinx, 5

Systema Naturae (Linnaeus), 20

T

television towers, 96

tencas, 49–53

Tennessee, 72

Tennessee Ornithological Society, 90

Texas, 72

Thompson, Maurice, 36

Three Stooges, 59

Titlow, Budd, 117

"To an English Nightingale," 36

To Kill a Mockingbird (Lee), 98,
99–101
Torfs, Ana, 18–19
tortoises, 51
turkeys, 8

U
University of Florida (UF)
research, 111–15

V
Visscher, J. Paul, 90

W
Washington, Mary, 27
Waterton, Charles, 44
Wayles, John, 25

What It's Like to Be a Bird
(Sibley), 8
Whistling Dick, 58
White House birds, 25–30
Williams, Alpheus S., 70
Wilson, Alexander, 39–41, 43, 76
Wilson, Woodrow, 80
Winner, Septimus, 57–60
Wisconsin, 106
Wright, Mabel Osgood, 81

X
X-chol-col-chek, legend of,
12–13

Y
Yaukey, Peter, 105